THE FIRST
BOOK OF FIRSTS

A *Golden Hands* book

Marshall Cavendish, London

Ski Club of Great Britain
E. Brenter
Kodak Museum
Post Office
David le Roi
Novosti Press Agency
The Science Museum of London
Victoria & Albert Museum
Illustrated London News
Royal Aeronautical Society
British Oxygen Company
United States Navy
Chard Jenkins
London Transport
Conway Picture Library
Ford of Britain
Bettmann Archive
Department of the Environment
Associated Press
Imperial War Museum
National Maritime Museum
Rothemsted Experimental Station
page 100: Major Wingfield from *Kings of the Court*
by E. C. Potter Jr. used by permission of Charles
Scribner's Sons. Copyright 1936 Charles Scribner's
Sons; renewal copyright © Edward C. Potter Jr.

Published by Marshall Cavendish Publications Ltd.,
58 Old Compton Street,
London W1V 5PA

©Marshall Cavendish Publications Ltd 1973
58 Old Compton Street,
London W1V 5PA

ISBN 0 85685 035 7

This volume first printed 1973

Printed by Colour Reproductions Ltd., Billericay, Essex,
and bound by R. J. Acford Ltd., Chichester, Sussex, England.

This edition is not to be sold in the U.S.A.,
Canada and the Philippines

A Book of Firsts

Don't take them all for granted – the bicycle and the light bulb, the pen and the bra, false teeth and Christmas cards, barbed wire and motor-racing.

Find out now. When was the *first?* The *first* vacuum cleaner, the *first* game of basketball, the *first* pillar box and the *first* tank?

What was the first stamp or the first postcard? What did the first motor-bike or the first tractor look like? What happened when the first submarine went to war or when the first parachute jump was made? What led up to the first radio or the first photograph?

Did you know that the first motor-driven dentist's drill was clock-work? Or that football first came to Britain with Caesar's Roman army? Or that in the first colour film every photograph on the reel was coloured by hand? Or that the hovercraft began with a cat-food tin, a coffee can and a vacuum cleaner?

Do you know what is the first recorded name of a human being?

Because if you don't this is the place to find out. And if you want to know exactly what many of these firsts looked like then this is the place to see them.

A first collection of firsts, from sport to medicine, from methods of travel to methods of communication, from household machines and war machines to agricultural machines and industrial machines – we've tracked them down to their earliest origins.

And if you know of any earlier ones we'll start preparing for another collection!

CONTENTS

COMMUNICATIONS

SPORT

AGRICULTURE

INDUSTRIAL

Part 1
AVIATION

Today jet aircraft flash across the sky, from one continent to another, in a matter of hours; space rockets streak towards distant planets; helicopters hover above patches in the forest; missiles can be fired from submarines at targets thousands of miles away – and can be intercepted by other missiles also thousands of miles away.

We have come a long way since Daedalus and his son Icarus escaped from the Minotaur's labyrinth on wings of feather and wax. Against his father's orders Icarus flew too near the sun. The wings melted and he fell to his death.

Since then hundreds of brave aeronauts have fallen to their deaths from a strange assortment of weird, wonderful and ingenious flying machines, some built out of little more than imagination . . .

OTTO LILIENTHAL. *One of the greatest pioneers of flying . . . killed soon after this picture was taken.*

The earliest balloons

IN THE MIDDLE AGES, the Chinese celebrated festivals and special ceremonies with small, open-necked balloons made of thin paper and filled with hot air. Weights were attached to the balloons to stop them overturning in gusts of wind and spilling out the air.

As long as the balloons did not burst into flames as they were being inflated, always a dangerous moment – they had to be held over a fire whilst the air inside them heated up – they would rise and float above the crowds until the air cooled.

The hot-air balloon

TWO FRENCH BROTHERS, Joseph and Etienne Montgolfier, were the first to make a serious and successful experiment with a hot-air balloon of any substantial size.

When they first began their experiments, hydrogen gas had just been discovered by the British scientist Cavendish. Hydrogen was only one-thirteenth of the weight of common air, and one or two of the brothers' rivals proved quickly enough that it could be used to lift a soap bubble or even the bladder of a calf. So why not a balloon?

But the Montgolfiers found that hydrogen also had its disadvantages. For one thing, it was highly inflammable; for another, they found it very difficult to produce a material for the balloon itself that would seal in the gas tightly enough. No balloon would stay long in the air if the gas escaped immediately, and hydrogen passed through paper as readily as through a sieve. Silk, a possible alternative, was equally useless.

Straw and wool

So the brothers turned to hot air, or smoke, which they believed was another kind of gas released by the burning of various materials. They were stubbornly convinced that different substances when burnt produced different qualities of gas, some lighter than others and therefore better suited than others to lifting a balloon.

For their first public demonstration, in the market place at Annonay, near Lyons, on June 4th, 1783, the Montgolfiers decided to use damp straw and chopped wool for their fire, having established to their own satisfaction that these produced

HYDROGEN LEADS AGAIN. *On December 1, 1783, Alexandre Charles once again fills his balloon – to make the first manned, hydrogen-balloon flight. Filling the balloon took Charles two days. The flight lasted eight minutes.*

the lightest possible gas. It was, in fact, nothing more than hot air but, as the Chinese had long ago proved, it worked.

The balloon was 38 feet in diameter, 35 feet high and weighed 430 pounds. It was loaded with an additional 400 pounds of ballast. The fabric of the balloon was linen, lined with paper.

Supported by a primitive scaffolding,

the balloon was retained by ropes hel by workmen from the village. The stra and rags were set alight; the balloon b came inflated by the hot air; the workme gripped hard against the upward strai When finally the ropes were released t balloon rose about 600 feet, though son reports say higher and some muc lower.

THE BALLOON THAT WAS "STABBED TO DEATH". *Alexandre Charles' un-manned hydrogen balloon landed in a village near Paris. The terrified country-folk thought it was a monster – and attacked it.*

THE MAN WHO THOUGHT OF HYDROGEN. *Alexandre Charles believed hydrogen was superior to heated air – and he was right. His first balloon reached 3,000 feet.*

1783. THE MONTGOLFIER BROTHERS *sent up the first hot-air balloon. They thought burning meat and wool produced a special "gas".*

The hydrogen balloon

THE TRIUMPH OF THE Montgolfiers with their hot-air balloon provoked the first successful launching of a full-scale hydrogen balloon the same year.

Outraged at the brothers' success, the eminent physicists of the Paris Academy, determined not to be outclassed by two provincials, encouraged one of the most promising of their younger members, Alexandre Charles, to make a practical experiment that would demonstrate the superiority of the Academy.

At first, Charles thought that the Montgolfiers had themselves used hydrogen. He was amazed when he learnt that this was not so. Hydrogen, he was convinced, would be a great deal more effective than 'Montgolfier gas' – hot air.

Astounded crowd

After frantic experiments and initial disasters in his back yard (it was a problem just to produce enough of the gas to fill the balloon), Charles made his first public demonstration of a hydrogen-filled balloon before the astounded crowd of fashionable Paris on August 27th, 1783, almost two months later than the Montgolfiers' experiment.

Soaring to a height that has been reported by one observer to be nearly 3,000 feet, the gravity-defying marvel disappeared over the Paris sky-line. Three-quarters of an hour later the balloon fell to earth in a field near the village of Gonesse, fifteen miles from Paris.

Terrified villagers

Charles assumed that it had burst under the pressure of the expanding gas. What remained when it landed was torn apart by the terrified villagers who, at the sight of the shapeless, billowing fabric settling in their field, believed that some dreadful monster was attacking them.

It had not been too difficult for Charles to overcome the problem that had deterred the Montgolfiers from using hydrogen. To stop the gas leaking through the balloon's skin, he had coated the skin with a rubber solution.

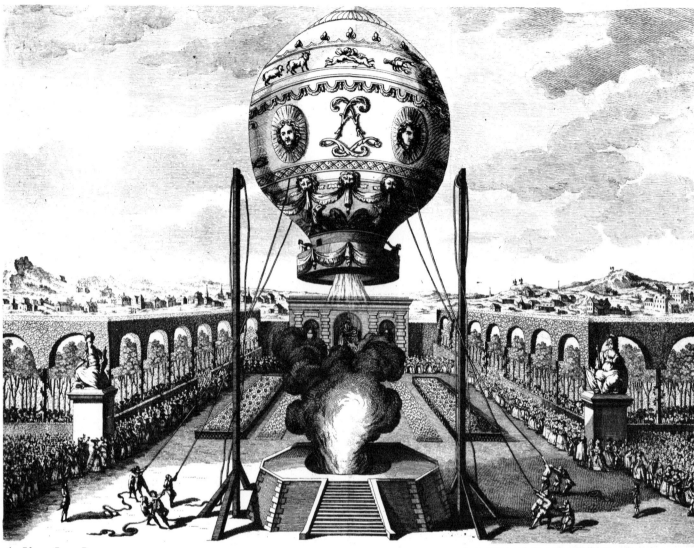

A live load

WITHIN A FEW DAYS of Charles's successful experiment with a hydrogen balloon, the Montgolfiers were back in the lead with the next step toward getting man himself into the air.

Their initial demonstration had aroused so much attention that they were asked to repeat it, first in Paris, before a committee of scientists, and then at Versailles, in the presence of Louis XVI and Marie Antoinette, his Queen.

To mark this great honour, the Montgolfiers provided an additional attraction. For the first time, a live load was to be raised in the balloon. The aerial passengers were a cock, a duck and a sheep.

The fuel – shoes

The first gun signal was at one o'clock, when the inflation of the balloon began. Eleven minutes later a second gunshot indicated that the balloon was fully inflated. (It had taken Charles *two days* to

THE FIRST MEN TO FLY. *Using a Montgolfier hot-air balloon, two French aristocrats, Pilatre de Rozier and the Marquis d'Arlandres, became the first men to fly.*

fill his hydrogen balloon!) At the third shot the balloon rose magnificently above the royal party and the crowd of an estimated 130,000 people. Eight minutes later the balloon – and the animals – came down unharmed.

Only one incident spoilt the occasion. Joseph Montgolfier's ideas about the different qualities of gas provided by the burning of different materials had become increasingly eccentric. One eye-witness at Versailles noted what happened:

'*They (the Montgolfiers) had caused all the old shoes that could be collected to be brought here and threw them into the damp straw that was burning, together with pieces of decomposed meat;*

for these are the substances which supply their gas. The King and Queen came up to examine the machine but the noxious smell thus produced obliged them to retire at once.'

A manned balloon

JUST OVER FOUR MONTHS after the Montgolfiers' initial triumph, another Montgolfier hot-air balloon became the first balloon to lift a man.

The King forbade the experiment on the grounds that it would be too dangerous, so it was proposed that criminals should be sent up first. Finally, it was an aristocrat, Pilatre de Rozier, who overcame the opposition of authority and his friends and insisted that he should make the courageous ascent.

On October 15th, 1783, de Rozier had his first tentative trial. The balloon carried its own fire. A fire basket made of wrought-iron wire was slung from the bottom of the neck of the balloon by chains. Portholes were provided in each side of the neck so that fuel could be fed to the fire from the gallery.

Five-mile flight

The balloon was tethered to the ground and did not rise more than 84 feet. But in later trials, also tethered, de Rozier demonstrated that he could control the rise and fall of the balloon to some extent either by adding more fuel or by damping down the fire.

Elated by his success, he persuaded his friend, the Marquis d'Arlandres, to accompany him on the first untethered flight over Paris. They ascended from a field outside the city on November 21st. The flight was low and carried them over five miles. It lasted twenty-five minutes.

A manned hydrogen balloon

BARELY A WEEK after de Rozier and d'Arlandres had proved that they could survive the unknown perils of flight, Alexandre Charles gave a similar demonstration with his balloon.

With his accomplice Ainé Robert, Charles made the first ascent in a hydrogen-filled balloon, from the garden of the Tuileries, on December 1st, 1783.

A channel-crossing by air

THE FIRST CROSSING of the English Channel was accomplished by a fiery-tempered Frenchman called Pierre Blanchard, over a year after his fellow-countryman, Pilatre de Rozier, had achieved the first manned flight.

Blanchard had been studying the flight of birds and had experimented with a weird machine in which two foot treadles and two hand levers were used to work four flapping wings. The pilot sat in a completely enclosed cockpit beneath the wings of this extraordinary 'ornithopter'.

Unfortunately the device never left the ground. But Blanchard was inspired by the success of the Montgolfiers and decided to add a balloon to help lift his machine. At the same time, he kept his flapping wings for propulsion.

No one in France paid much attention to Blanchard and his absurd-looking contraption, so he went to England, where the competition was not so fierce. There he had two successful flights late in 1784 and persuaded a wealthy American, Dr John Jeffries, noted for his interest in meteorology, to pay all the expenses for a third trip on January 7th, 1785.

The cost of the expedition was over £700 – quite a sum.

The launch took place from Dover Cliff at one o'clock in the afternoon. The aim was to cross the Channel. But the balloon was loaded down with a great weight of useless gear including the ridiculous wings, a rudder and a *moulinet* (rotating fan intended to propel the balloon).

Jeffries described what happened: 'When two-thirds over we had expended the whole of our ballast. At about five or six miles from the French coast we were again falling rapidly towards the sea . . . we cast one wing, then the other; after which I was obliged to unscrew and cast away our *moulinet*; yet still approaching the sea very fast, and the boats being much alarmed for us, we cast away, first one anchor, then the other, after which my little hero (Blanchard) stripped and threw away his coat (great coat). On this I was compelled to follow his example. He next cast away his trousers. We put on our cork jackets and were, God knows how, as merry as grigs to think how we should splatter in the water.'

No anchor

Nonetheless, they passed over the French coast safely at exactly three o'clock and came down twelve miles inland, 'in the midst of the forest De Felmores, almost as naked as the trees, not an inch of cord or rope left, no anchor or anything to hold us, nor a being within several miles'.

It was a successful end to man's first overseas flight, because two of the most important items of equipment were still on board: a bottle of brandy and a packet of letters – the first airmail letters.

RIGHT: THE FIRST OVERSEAS FLIGHT.
1785. Pierre Blanchard succeeded in crossing the Channel in this strange aircraft – but not the way he had expected.

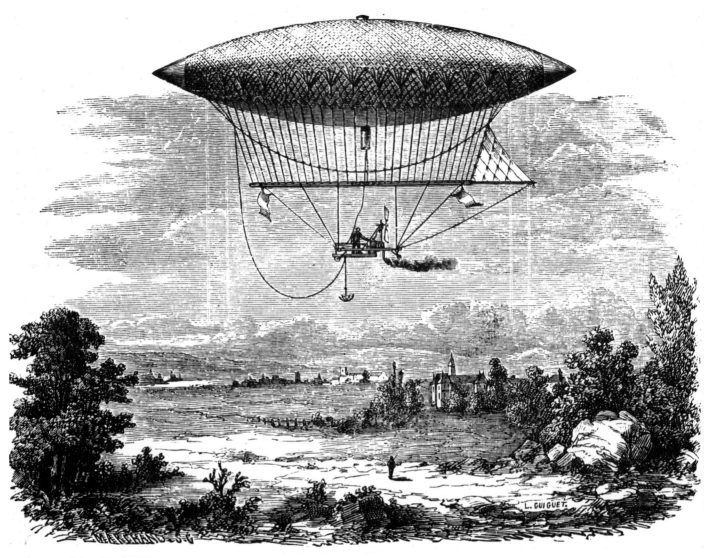

Impractical airships

THE EARLIEST AIRSHIPS were mostly fanciful and impractical. A typical example was that designed by an Italian scientist called Francisco Lana over a hundred years before the Montgolfiers' balloon ascent. It may not be the first impractical design but it is a very good example of the sort of weird contraption scientists thought up in those days.

Lana published a book in which he included two chapters on the 'aerial ship'. His invention was to be buoyed up by suspension from four globes made of thin copper sheeting, each about 25 feet in diameter.

In order that the globes should become lighter than the air outside them and so be able to support the weight of two or three men, Lana proposed that the air from within each globe should be pumped out. What he did not take into account was the effect that atmospheric pressure

A STEAM ENGINE TAKES THE AIR.
September, 1852. The first steam-driven airship soars over, and away from, Paris. It carried 5 cwt of coke and water.

would have on the globes: with a vacuum inside them, they would probably have caved in.

No city safe

In any case, Lana's proposed method for propelling his airship once it was in the air was equally impractical. There was a false idea that had persisted in the minds of the early experimenters for a long time that what worked on the sea would work as well in the air. Lana's airship was to be equipped with oars and sails!

In his book he saw only one real problem in the development of the airship: 'I do not foresee any other difficulties that

prevail against the invention save one,' he wrote, 'which seems to me the greatest of them all, and that is that God would never surely allow such a machine to be successful since it would create many disturbances in the civil and political governments of mankind. Where is the man that can fail to see that no city would be proof against surprise when the ship could at any time be steered over its squares, or even over courtyards of dwelling-houses, and brought to earth for the landing of its crews? Iron weights could be hurled to wreck ships at sea, or they could be set on fire by fireballs and bombs, nor ships alone but houses, fortresses and cities could thus be destroyed with the certainty that the airship could come to no harm, as the missiles could be hurled from a great height.'

Lana anticipated the destructive power of the airship correctly but he was greatly mistaken in thinking that it would not be developed for that reason.

Disaster . . . for the first aluminium airship

Design for an airship

ONCE THE MONTGOLFIERS had proved it possible for man to ascend in a balloon, the interest in designing one that could be steered and powered and made to move against the wind rather than to drift helplessly before it grew more keen.

An airship, or 'dirigible' (*the name came from the Latin word meaning 'to direct'*) was the next logical step forward in man's conquest of the air. And the first practical design to appear was drawn up in 1784 by Jean Baptiste Marie Meusnier.

Meusnier was a soldier. He was equally distinguished as an engineer and a scientist. Even though his design, like that of Lana's far stranger invention, was never executed, it is Meusnier who is thought of as the father of the dirigible balloon.

Hand bellows

His greatest contribution to the development of the balloon was the idea of using a 'ballonet', or 'little balloon', an auxiliary reservoir inside the main envelope (as the balloon itself was called) which could be filled with air from the car of the balloon by hand bellows. This gave the aeronaut better control over the lift of the balloon. In later developments the bellows were replaced by a powered compressor.

Meusnier's drawings show an envelope in the form of an ellipsoid – an oval shape – 260 feet long with a capacity of 60,000 cubic feet. This envelope had a strong 'belly-band' of fabric to which the ropes supporting the car were attached. Between these car ropes there were flexible pipes connecting the bellows on the car to the ballonets.

Designed to float

The car itself was a long, slim, boat-shaped hull. Meusnier did not intend to take any risks: the car was designed to float in case he was forced down onto a stretch of water. It was even equipped with a rudder – but no oars and no sails! Instead there were three large propellers mounted on a common axis in a framework between the car and the envelope. It was intended that these propellers should be driven manually from the car by means of ropes and a pulley.

A successful airship

IT WAS NOT UNTIL 1852, nearly 70 years after the first balloon ascent, that Henri Giffard, a French engineer (the French seem to have led the way at first), built the first working dirigible.

His spindle-shaped airship had a capacity of 88,300 cubic feet, nearly half as large again as Meusnier's earlier design. Over the envelope was stretched a net from which, by means of ropes, a heavy pole, 66 feet long, was carried beneath the envelope in a horizontal position. This can be seen clearly in the illustration. At the end of this pole there was a rudder in the form of a triangular sail.

The car was suspended from the pole. It was slung 20 feet below and contained the engine and the propeller. It was important that the engine should be as far below the envelope as possible in order to avoid igniting the hydrogen with which the envelope was filled.

Giffard had designed the engine himself. It was driven by steam and developed three horsepower. The propeller had three blades and was 11 feet in diameter. It spun at 110 revolutions a minute.

The engine and boiler together weighed 350 pounds. With Giffard on board the total weight that had to be lifted was one and a half tons – light enough for him to carry five hundredweight of coke and water.

Dressed very correctly, in a top hat and frock coat, Giffard rose majestically in his steam balloon from the Paris Hippodrome on the 24th September, 1852. To the amazement of a large crowd he flew slowly out of sight, leaving behind a white trail of exhaust steam.

Henri Giffard travelled 17 miles at a steady six miles an hour and made a safe landing. It was only a small beginning. He had chosen a day of perfect calm, which was comparatively rare. Under average weather conditions his little airship would have been almost as helpless as an unpowered balloon.

An all-metal airship

THE AUSTRIAN ENGINEER who designed the first aluminium rigid airship never lived to see it in the air. This may have been a good thing. The experiment came to a disastrous end and was incompetently handled.

David Schwartz started designing his airship in 1895. Construction began in Berlin but the dirigible was not completed until two years later, by which time Schwartz had died, leaving his widow in charge of the project.

1897. A DREAM ENDS. *David Schwartz's all-aluminium airship crashes at Berlin, after breaking loose before take-off.*
But, in the tragedy, was one element of success. The new rigid frame survived the crash.

The first man drops from the sky

The envelope of the Schwartz airship consisted of a tubular aluminium framework covered with aluminium sheeting eight thousandths of an inch thick. This envelope had a conical nose, like a rocket, and a slightly concave stern. The overall length was 156 feet; its capacity was 130,000 cubic feet.

Three propellers

The car was slung beneath the envelope, just as Giffard's had been, but this time it was attached by aluminium struts. There were also two aluminium tractor propellers mounted abreast on either side of the bow of the car. Above the stern was a third 'pushing' propeller, which was movable and could be used for controlling the direction of the airship.

All three propellers were driven by belts at 480 r.p.m. from a four cylinder Daimler petrol engine which generated 12 horsepower – four times as powerful as Giffard's engine.

The filling of the airship was carried out on the Tempelhofer Field, Berlin, during the first two days of November, 1897, under the direction of Captain von Sigsfeld. However, there seems to be a good deal of disagreement as to who was the pilot.

A young soldier

One account says that the pilot was Herr Jaegels, Schwartz's chief engineer; according to another it was a young mechanic called Platz; a third alternative was a young soldier from the balloon corps. Whoever it was could not have had very much aeronautical experience; certainly, he had no experience at all that could have helped him in controlling single-handed Schwartz's novel and complicated machine. The first flight, on November 3rd, was chaotic.

Giffard had been lucky enough to have a mild day for his first powered flight; this time there was a strong wind blowing at 19 miles an hour. One by one the ropes that held the airship to the ground broke under the strain. With his engine running, the alarmed pilot tried to increase speed in order to clear the ground. His right hand propeller belt slipped off its wheel; then the band of the rudder screw also came off.

LEFT: THE FIRST PARACHUTIST. *Andre Jaques Garnerin swings down towards Grosvenor Square, London in 1802. This "drop" was 8,000 feet. Garnerin used a balloon to take him up in his parachute-car. Then he cut himself loose.*

Out of control

By now the airship was moving forward, just above the ground, out of control. Confused and terrified, the pilot opened the valve of his engine in order to land as soon as possible; at the same time he attempted to check the force of his descent by throwing out ballast.

The airship bounced three times before it was thrown on its side on the slope of a hill. Only the framework remained intact. At least that proved its strength. The rest was reduced to a wreck, partly by the wind and partly by the curious crowd which gathered round and pulled it apart.

The trial could not honestly be called a success. The real success lay in the survival of the rigid frame. This proved that it was possible to construct a riveted metal balloon that really could lift off and drive itself forward against a breeze – even if only for a very short time!

The earliest parachutes

THE IDEA OF A "FLYING MAN". *As early as 1840, the idea of parachuting had come to Fausto Veranzio in Venice. This is his own drawing.*

UNTIL RECENTLY it has been generally believed that the first parachute was designed and sketched by Leonardo da Vinci. His tent-like sail was drawn in about 1485. But a manuscript in the British Museum has two other drawings, possibly done by an engineer from Sienna, a good five years earlier than Leonardo's.

It was not until 110 years after Leonardo's sketch that the first published illustration of a parachute appeared. This was in a book called *Machinae Novae* (New Machines), published in Venice by Fausto Veranzio, in which he described and illustrated a *homo volans* – a 'flying man'.

Veranzio's inspiration clearly came from the sail of a ship. It is interesting to note that neither the sunshade nor the umbrella – on whose shape the modern parachute is based – had any influence over the very earliest designs.

A human parachute drop

WHAT WAS REALLY NEEDED before the parachute was developed was the development of a balloon from which to drop it. So it was not until 14 years after the Montgolfiers' first successful flight that Andre Jaques Garnerin hitched a tiny car and furled parachute to a hydrogen balloon and bravely ascended into the air above the park of Monceau, in France.

Garnerin was 27 when he made this attempt. He had always been a keen aeronaut but the French Revolution of 1789 had interrupted his flying career. A few years later he joined the army. He was taken prisoner by the British, handed over to the Austrians, and imprisoned in a fortress in Buda, in Hungary, for nearly three years.

During this time he dreamed of constructing a parachute with the help of which he could escape. The scheme came to nothing but the idea lingered on, ready to be put into practice when he returned to Paris.

Cut the cord

The parachute that Garnerin used on 22nd October, 1797, was ribbed and folded like a modern sunshade. When his balloon had flown high enough over the park – about 3,000 feet – he cut the cord that attached his car to the balloon. The balloon shot upwards and burst.

The car and parachute dropped downwards abruptly. But, as Garnerin fell, his canopy, 30 feet in diameter (when laid out flat) and made of canvas, opened perfectly in the rush of air.

The descent, however, was uncomfortable. The fabric was too thick to spill any of the wind. This meant that the parachute came down in a series of oscillations, swinging violently like a pendulum, with Garnerin clinging tightly to his little basket in order not to be thrown out.

Safe to earth

He landed safely, if roughly, on the plain of Monceau, and rode back on a horse in triumph to the park. The drop was described by a contemporary as 'one of the greatest acts of heroism in human history.'

THE FIRST AEROPLANE TAKES OFF. *December 17, 1903. Orville Wright lies flat on the wing during take-off, watched tensely by his brother, Wilbur. Wilbur has just let go of the wing. The plane is taking off from the end of a 40-foot wooden rail. It landed on skids.*

A piloted and controlled glider

THE FIRST proper manned glider to achieve any reasonable kind of flight was built by a man called Cayley, in 1853. But Cayley's pilot did not steer the glider, and the controls were locked.

Otto Lilienthal's gliders, built between 1891 and 1896, were the first to be controlled by their pilot, who was Lilienthal himself, one of the greatest men in the history of flying. It was his mastery of controlling gliders in flight that made it possible for the Wright brothers to achieve their success with a powered machine several years later.

Lilienthal was killed in a flying accident, but if he had survived he might well have succeeded in making a powered flight before the Wright brothers.

The fatal wind

His initial attempts were made from a springboard at his home in Germany. His first recognisably successful glider was built in 1893. This was his Number Three, that had a wing-area of 107.5 square feet and a fixed tailplane. Like all his gliders, it was a 'hang-glider' – the pilot hung from the frame by his arms and could swing his body and legs in any direction in order to shift the centre of gravity and so achieve a limited control in the pitch and roll of the glider.

He crashed on August 9th, 1896, whilst gliding in one of his standard Number Eleven monoplanes (he also built several biplanes). A gust of wind brought him to a standstill in mid-air. He hovered there for a second, then threw himself forward in order to get the nose of the glider down before he stalled. But the starboard wing dropped and the glider side-slipped to the ground, crumpling its wing but leaving the rest of the structure undamaged.

Lilienthal's spine was broken. He died the next day. He himself had said many times: 'Sacrifices must be made'.

Powered flight in an aeroplane

THE FLIGHT THAT Orville Wright made on December 17th, 1903, is one of the most famous moments in aviation history. The powered aeroplane was at last proved possible.

Orville and Wilbur Wright first became interested in aeroplanes when their father bought them a present of a toy helicopter, constructed of cork, bamboo and thin paper, driven by twisted rubber bands. They tried to build larger versions of the toy. A few years later Orville started building kites – so successfully that he sold several to his schoolfriends.

After leaving school, the brothers teamed up and worked together on a number of engineering projects. They were athletic and became enthusiastic cyclists. They opened their own cycle shop and, in 1895, began to manufacture bicycles themselves. One of these was a low-priced machine that carried their own name: the 'Wright Special'.

Face downward

Meanwhile their interest in flying never faded. They studied every possible article on aviation. Lilienthal's experiments impressed them enormously and even when he was killed they pressed ahead with their own experiments.

To convert their glider into an aeroplane, it needed an engine. But no firm could supply them with a petrol motor light enough to meet their requirements. So they built their own. It weighed only 170 pounds and developed 12 horsepower (the same as David Schwartz's first all-metal airship).

When completed, the Wright brothers' aeroplane had a wing span of 40 feet, or just over. The upper and lower wings were six feet apart. The engine was fixed to the lower wing, a little to the right, so that it balanced the weight of the pilot, who lay face downwards in order to reduce wind resistance.

There were two 'pusher' propellers, which revolved in opposite directions, so that there was no risk that their motion would make the aeroplane itself start twisting round. A rudder at the tail consisted of two vertical vanes. The bi-plane elevator stretched in front of the wings.

40-foot run

Performance was recorded by a stop watch, an anemometer (an instrument for measuring the force of the wind) and a gadget to count the number of revolutions made by the motor. The aeroplane was set up at the end of a 60-foot-long wooden rail, from which it took off. It could land anywhere by means of its skids.

Orville described the first attempt: 'After running the motor for a few minutes to heat it up, I released the wire that held the machine to the track, and the machine started into the wind. Wilbur ran at the side of the machine, holding the wing to balance it on the track. . . . The machine, facing a 27 miles-an-hour wind, started very slowly. Wilbur was able to stay with it until it lifted from the track, after a 40 feet run. . . .'

Orville does not sound very excited, but it was the moment towards which all their work and their hopes had been building up. The flight lasted only twelve seconds. But it succeeded, and that was what mattered.

RIGHT: OTTO LILIENTHAL. *The dare-devil glider-flying of Otto Lilienthal made it possible for the Wright brothers later to succeed with powered flight. Here, Lilienthal is pictured a few weeks before his death crash.*

1915. HUBERT JUNKERS' ALL-METAL, CANTILEVER MONOPLANE. *Though the German authorities said it was too heavy, it reached 105 m.p.h. This was the only one built but the "tin donkey", as it was called, influenced the next years of aviation design.*

DEVON CREAM, AND LETTERS, FOR PARIS. *August, 1919. The converted de Havilland bomber which flew the world's first scheduled air service – a 2½-hour trip to Paris.*

THE BOMBERS THAT BEGAN AIR SERVICES. *This is the de Havilland bomber, one of the squadron used to fly British politicians to the Paris Peace Conference in 1919.*

THE FIRST OF THE SUPERSONICS. *1947. The first supersonic rocket piloted plane, the American Bell XS-1. Its first record-breaking flight was 670 m.p.h.*

The all-metal aeroplane

In 1910, a German, Hugo Junkers, patented his latest invention – a thick-section cantilever wing. This means that the wing was supported at one end only, without any external bracing. But it was not until five years later that the invention was applied successfully in aviation.

The J.1. (Junkers Mark 1 – it was also called the 'tin donkey') was the first all-metal (iron and steel) fully cantilever monoplane. Only one was built. Its first flight was on December 12th, 1915. It was powered by a 120 horsepower engine and had a top speed of 105.5 miles an hour.

The German authorities criticised the 'tin donkey'. They said that the steel structure was too heavy. But Junkers ignored them and continued to improve his machines, which had a great deal of influence on later aeroplanes.

A manned supersonic rocket aeroplane

The first manned supersonic rocket aeroplane was the American Bell XS-1. Its first powered flight was on December 8th, 1946. It was launched in the air from a Boeing B-29 Superfortress. Its first supersonic flight was made on 14th October, 1947, at 670 m.p.h. at 42,000 feet.

A regular international air service

THE R.A.F. established the first regular air service to places outside Britain. In December, 1918, a communications squadron was formed to carry members of the British Government to Paris for the Peace Conference following the First World War. A second unit carried airmail to Cologne for the Rhine army of Occupation.

However, the world's first regular international scheduled civil air service began on August 25th, 1919, when a de Havilland 4A converted day-bomber left Hounslow for Le Bourget, Paris, where it arrived two and a half hours later.

On board were one passenger, small consignments of leather, some Devonshire Cream, newspapers and mail, together with several brace of grouse.

The operator was Aircraft Transport and Travel Ltd. Single fare to Paris – £21.

Before taking off on these early passenger flights, everyone was issued with a heavy leather jacket, goggles and a helmet. At first passengers used to take their outfits home as souvenirs.

The adventure was so alarming that last minute nerves were common. There was an official 'passenger encourager' at Hounslow airport who dressed the passengers for the flights and served stiff drinks to the very nervous, 'on the house'.

The earliest helicopters

THE EARLIEST HELICOPTER was a toy, which existed even in the fourteenth century. The toy consisted of four blades mounted on a spindle and dropped into a holder. A long string was wound round the spindle and taken out through a hole in the holder. When the string was pulled, the spindle rotated rapidly and the toy went spinning up into the air. Both the spindle and the holder were made of wood.

The idea of the helicopter came from the windmill. The blades of a windmill form what is called a 'passive airscrew'. This means that they remain stationary and are turned by the wind. The helicopter does exactly the opposite: the blades of the propeller are turned by their own power and so draw the aircraft through the air. This, of course, is what the ordinary aircraft propeller does as well.

A manned helicopter

IN 1907, two full-sized machines succeeded in rising vertically from the ground. The first, built by Louis Brequet and Professor Richet, was tethered. It rose about two feet off the ground. At a later attempt, it managed to rise five feet.

Two months later, in November, near Lisieum, in France, Paul Cornu achieved the first free helicopter flight. It was an impressive success even if it was not a very great one: he rose only about one foot into the air, and hovered there for only about 20 seconds.

A practical production helicopter

THE MAN WHO FINALLY DEVELOPED the helicopter into a practicable and controllable aircraft, and produced a design that could be built in quantity, was Igor Sikorsky.

Sikorsky designed and built a helicopter in 1909, and another the following year. Both of these were unsuccessful. He left Russia during the Revolution and went first to France and then to America, where he founded his own company with the help of friends and continued with his designs.

It was not until 30 years after building his first helicopter that he flew the successful VS300, on September 14th, 1939. During the initial flights it was tethered to the ground, but later it made free flights.

Sikorsky made continuous tests on this design for three years, tests that led eventually to the R-4, the first production helicopter.

THE PLANE SIKORSKY BUILT. *1942. Helicopters that could be mass-produced – the 30-year brain-child of Russian-born Igor Sikorsky.*

Part 2

ROAD and RAIL

Of course it all began with the wheel – probably the greatest break-through man has ever made. And then, for thousands of years, there were chariots and carriages and coaches and carts and wagons and every conceivable shape and size of vehicle that could be put on wheels and drawn by people or horses or bullocks or elephants or whatever kind of animal was available.

Until new sources of power were discovered. First came the steam engine, then the internal-combustion engine, then electricity – so at last there were alternatives to animal power, machines that could bear heavier loads faster, even if at first they did break down as often as animals fell sick.

A whole new world opened up, of engines and railways, of tunnels and cars. Distances became almost insignificant. Places that for as long as anyone could remember had been so many days' ride away suddenly took only a few hours to reach. Whether people liked it or not, everywhere seemed nearer.

THE FIRST CIRCULAR RAILWAY. *London, 1808. An exhibition staged by inventor Richard Trevithick. Price of a ride . . . one shilling.*

THE FIRST BICYCLE *was adapted from the idea of a child's hobby horse. A German baron added the moveable steering wheel. A speed of eight, even ten miles an hour could be reached.*

The bicycle

THE DASHING COMTE DE SIVRAC caused great amusement in the gardens of the Palais Royal, in Paris, when he started a new fashion by making the first known appearance on a bicycle in 1791. He appeared in the park sitting astride a small wooden horse, fitted with two wheels, which he pushed along by thrusting at the ground with each foot in turn.

This kind of vehicle had probably existed for some time as a children's toy. Never before had a respectable gentleman been seen on one. But the fashion did not last long, and no improvements were made to de Sivrac's wooden horse for another 28 years.

The first practical bicycle was invented by an eccentric German engineer who

LEFT: THE BICYCLE GETS ITS PEDALS. *Kirkpatrick Macmillan invented the first pedal and-rod bicycle. Note the horse's head . . . derived from the hobby horse.*

was also an expert on agriculture – the Baron von Drais de Sauerbrun. He revived the fashion that de Sivrac had begun and adapted it to his own needs in 1819.

Rough country

The Baron's work involved long journeys over steep and rough country. By adding a comfortable saddle, some handlebars and a steerable front wheel to de Sivrac's frame, von Drais created a very useful method of transport – at least along the flat and downhill. For the first time the bicycle became something more than a frivolous novelty.

The 'Draisienne', as it was called, appears to have made a very good reputation for itself. In the same year that it was first introduced, Ackerman's *Repository of Arts and Sciences* reported that, *'for such as take exercise in parks, or who have an opportunity of travelling on level roads, these machines are said to be very beneficial. A person who has made himself tolerably well acquainted with the management of one can, without difficulty, urge himself forward at the rate of eight, nine or even ten miles an hour ...'*

Von Drais himself never made much money out of his invention, even though the popularity of the 'Draisienne' quickly spread through France to England and the United States. Cycling – of a sort – once again became a fashionable sport.

The pedal bicycle

THE BICYCLE as we know it today was invented twenty years later than the 'Draisienne', in 1839, by Kirkpatrick Macmillan, a blacksmith from Courthill, Dumfriesshire.

The Macmillan bicycle had wooden wheels that were shod with iron tires – a rough ride! Its frame consisted of a curved wooden backbone, forked to hold a rear driving wheel. The front steering wheel was also held in an iron fork. The handlebars, just as on a modern bicycle, were attached to the pivot of this fork, which passed up through the frame.

No chain

There was no chain. The pedals were connected by rods to cranks that turned the rear wheel. Of course, once the rider's feet were off the ground and busy pedalling, it was necessary to learn to keep one's balance – something the cyclist had not had to worry too much about on the 'Draisienne'. Once he'd learnt that skill, he was more or less in control of the first real bicycle.

Macmillan himself rode one of his own machines for many years and for quite long journeys – some, obviously, too long: in 1842 he was fined five shillings at the Gorbals Police Court for knocking down a child at the end of a 40-mile ride from Courthill to Glasgow!

The railway track

IT IS NOT KNOWN when the first railway track was built. The earliest wheels gouged out ruts in the roadway and in some of the earliest pavements primeval engineers made stone ruts deliberately.

The Greeks made stoneways in this manner for the transport of heavy materials, and there are grooved stone tracks in the ruins of Pompeii. Wooden rails were known in England in the reign of Elizabeth 1, and in the seventeenth century they were being used in mining districts.

Wooden trucks

As far as we know, the first proper rail track to be installed in England was by Huntingdon Beaumont, in 1604, at Wollaton Colliery, near Nottingham. Four years later, Beaumont left Nottingham and installed rail tracks at mines in the north, so that by the end of the century railways were a common feature of mining and quarrying districts.

The picture of Prior Park is one of the earliest pictorial evidences of the use of rails for the transport of heavy materials in England. Prior Park is near Bath and was owned at the time by Ralph Allen. The wooden trucks were used to carry stone from the local quarries to barges moored on the River Avon. The wagon way also had wooden rails and was made wide enough for pedestrians and other traffic to use as well.

WHEELS ON RAILS. *One of the first railways, used to carry stone from quarries to barges on the River Avon. The rails were wooden. The rail "way" was made wide so that pedestrians and other traffic could move alongside.*

A passenger-carrying steam carriage

RICHARD TREVITHICK, a Cornish mining engineer, was the first to test a 'travelling engine'. Other inventors had refused to experiment with high pressure steam because they thought that it was too dangerous to handle. Trevithick was not put off by this. He tried out his ideas first of all on stationary engines and from the knowledge he gained from these experiments he constructed the first passenger-carrying steam carriage.

The first test was on Christmas Eve, 1801, up Beacon Hill, Camborne. The engine started well enough, until too many curious sightseers clambered on. Overloaded, it came to a stubborn halt. The steam pressure was not enough to draw the heavy load.

Six miles an hour

Three days later, Trevithick had more success – until, this time, the engine turned over in a ditch. Shaken by the accident, Trevithick and his friends 'adjourned to the hotel and comforted their hearts with a roast goose and proper drinks, when, forgetful of the engine, its water boiled away, the iron became red hot, and nothing that was combustible remained of the engine.'

But Trevithick was not discouraged. A year later, he went to London to patent his engine and, in the following year, he tried out an improved version of the carriage on the streets of London. With several passengers on board, the carriage travelled at about six miles an hour.

This engine had a horizontal cylinder, from which the exhaust steam was passed into a chimney. Bellows supplied a forced draught to the boiler. It was necessary for the coach body to be mounted high so that it rode above the large driving gears.

Unfortunately, many people regarded the carriage with suspicion and most people took no notice of it. It was soon broken up. If the public had been more interested in mechanical road transport and if the general road conditions had been better, it might have been more successful.

A steam engine on rails

AFTER FAILING to interest the public in his steam carriage, Richard Trevithick turned his attention to the railways. The first stream engine to run on rails was the result of a bet.

A South Wales ironmaster, Samuel Homfray, had been one of the few people to show any enthusiasm for the original 'travelling engine'. He laid a bet with another manufacturer that ten tons of iron could be hauled by steam along the tramway from the Penydaren iron-works to Abercynon on the Canal.

The trial was carried out on February 11th, 1804. Trevithick wrote: *'We carried ten tons of iron, five wagons, and seventy men riding on them . . . It's about nine miles which we performed in four hours five minutes, but we had to cut down some trees and remove some large rocks out of the road. The engine . . . went nearly five miles per hour . . . The gentleman that bet five hundred guineas against it rid the whole of the journey with us and is satisfyde that he has lost the bet . . . The publick until now called me a scheming fellow but now their tone is much alter'd!'*

Later the same locomotive hauled 25 tons at four miles an hour. But that was too ambitious a load for the tramways that were too weak and brittle to carry the weight. Eventually the engine was dismantled and used for stationary power only.

A circular railway

THE FIRST CIRCULAR RAILWAY was opened in London in July, 1808. The engine was called *'Catch me who can'*.

After helping Samuel Homfray win his bet of 500 guineas by carrying ten tons of iron on a steam engine, Trevithick decided to take one of his engines to London again for a series of demonstrations that would – he hoped – arouse public interest and enthusiasm. He was confident that he would win recognition in the end. Once again, he was disappointed.

Too heavy

An engineer described what went wrong, in a letter: 'I rode with my watch in hand at the rate of 12 miles an hour', he wrote. 'Mr Trevithick then gave his opinion that it (*'Catch me who can'*) would go 20 miles an hour, or more, on a straight railway; the engine was exhibited at one shilling admittance, including a ride for the few who were not too timid. It ran for some weeks, when a rail broke and occasioned the engine to fly off on a tangent and overturn . . . Mr Trevithick having expended all his means in erecting the works and enclosure, and the shillings not having come in fast enough to pay current expenses, the engine was not set again on the rail.'

Trevithick's locomotives failed mainly because they were too heavy for the types of rail then in use and partly because of the public's unenthusiastic attitude towards them. But his experiments were the basis and the inspiration for later railway pioneers.

LEFT: WHEELS WITH STEAM. *1804. Richard Trevithick leads the first steam engine on rails on its nine-mile run. Time: 4 hours 5 minutes.*

RIGHT: STEPHENSON'S 12 m.p.h. ENGINE. *The "Locomotion", built by George Stephenson, the first steam engine to be used on a public railway. Speed: 12 m.p.h.*

THE RAILWAY AGE BEGINS. *An artist's impression of the opening of the first public railway between Stockton and Darlington, England, in 1825.*

A public steam-operated railway

THE FIRST RAIL of the Stockton and Darlington Railway was laid on May 23rd, 1822, and on 27th September, 1825, the line was opened with much publicity.

The man who had perfected the locomotive and whose name is perhaps the best known in railway history was George Stephenson, who had made his reputation as an engine-wright at the High Pit Colliery near Killingworth, where he was given a free hand to construct a line on which the loaded wagons could be hauled by cables worked by stationary engines.

Stephenson became an acknowledged expert on railway construction and was the obvious choice for the post of surveyor and engineer of the proposed Stockton-Darlington line. The line was intended to join the coal mines west of Darlington to the docks at Stockton-on-Tees and, although it was principally intended for the transport of coal, it was to be a public railway, unlike earlier schemes.

There was a lot of opposition to the railway from local landowners – just as there is today to any new road plans – and also from those who operated stage coach services and who feared the competition. Even Parliament first rejected the plan: it was feared that the line would interfere with fox-hunting.

When the line opened eventually, 'the signal being given', wrote a local newspaper reporter, 'the engine started off with this immense train of carriages; and such was its velocity that in some parts the speed was frequently 12 miles an hour'. As it neared Stockton, many of the excited onlookers leapt on to the train, clinging to the sides, until by the time it arrived, there were almost 600 passengers on board.

The engine, called *Locomotion,* later surprised everyone by racing a stage coach between Darlington and Stockton – and winning! News of this triumph greatly increased the popularity of the new line, which was far more of a success than anyone had anticipated. As a result, Stephenson designed a special passenger coach, called the *Experiment,* to replace the open trucks that had been used at first.

Topsy-turvy

The railroad soon spread to America. A New York paper foresaw: 'What would be the effect of the railroad system – it would set all the world a-gadding. Twenty miles an hour! Why you will not be able to keep an apprentice boy at work, every Saturday evening he must take a trip to Ohio to spend the Sabbath with his sweetheart. Grave, plodding citizens will be flying about like comets. All local attachments must be at an end. It will encourage flightiness of intellect. Veracious people will turn into the most immeasurable liars, all their conceptions will be exaggerated by their magnificent

RIGHT: SHARE-HOLDERS IN CONVOY. *1863 . . . the future P.M., Mr Gladstone, and his wife (both arrowed) join shareholders in a test trip of the London Underground.*

notions of distance. Upon the whole, Sir, it is a pestilential, topsy-turvy, harum-scarum whirligig . . . I go for beasts of burden; it is more primitive and Scriptural and suits a moral and religious people better.'

An old lady, who, on seeing a railway for the first time, described 'a long black thing, spitting out smoke and crawling along the ground' which finally, on seeing her, 'uttered a loud yell, and rushed into a hole in the ground'.

A Thames tunnel

THE FIRST TUNNEL beneath the Thames connected Rotherhithe on the south bank with Wapping on the north. It was opened in March, 1843.

To construct the tunnel, the engineer, the famous Sir Marc Brunel, used a rectangular iron shield shaped like a huge box with open ends and projecting teeth. This was placed against the working face of the tunnel and moved forward as the excavation progressed, protecting the men inside from the immense pressure of the earth and water above. Behind the miners came the bricklayers, to line the tunnel.

Setbacks

It was originally intended that the tunnel should lessen the work of the London docks and save horse-drawn traffic from making the long detour by London Bridge. But long sloping approaches were needed for this purpose and by the time the actual tunnel was completed the company formed to construct it had, not for the first time, run out of money. They were forced to economise by providing access to the tunnel by staircases down steep shafts at either end. As a result, only pedestrians could make use of the short cut.

There were plenty of other setbacks, including a seven year period when, because of lack of funds, the project had to be temporarily abandoned, and an accident when the Thames burst through the roof drowning seven men. It was hardly surprising that the unlucky tunnel became known as 'Brunel's White Elephant'.

The London Underground

THE METROPOLITAN LINE between Farringdon Street and Bishop's Road, Paddington, was opened to the public on January 10th, 1863. The public rode in closed carriages. But, a few days before, the Prime Minister, Mr Gladstone, and his wife had ridden through the echoing tunnel in open trucks. A band greeted the train on its arrival at Farringdon Street and several hundred people were invited to attend a great banquet at the station to mark the opening.

The man who put forward the original idea for the Underground died the year before it was completed. Charles Pearson, a City Solicitor, was interested in penal reform and the embankment of the Thames; he was also disturbed at the congestion of London's streets, and he searched for a way to relieve some of the chaos.

Pearson's plan

Traffic problems are not exclusive to our own age. Even half way through the last century it has been estimated that 75,000 people entered London every day for work. The streets became blocked with a variety of vehicles – omnibuses, hackney carriages, coaches – and their iron wheels banged over the cobbles adding increasing noise to the confusion.

Pearson planned for a steam-operated railway slightly over three miles long, underneath Farringdon Road and King's Cross Road to King's Cross Station and then beneath Euston Road, Marylebone Road and Praed Street to Paddington. It would serve as a link between three of the main-line termini: The Great Western at Paddington, The London and North-Western at Euston and The Great Northern at King's Cross. Farringdon Street was chosen as the site for the eastern terminus largely because the

City Cattle Market, then occupying the area, was about to be moved to Islington.

Deep trenches

Construction of the tunnel began in 1860. The method used was called 'cut and cover'. Deep trenches were dug along the streets, lined with bricks and covered with girders or brick arches. The road was then restored over the tunnel. Although many of the householders along the route claimed that the digging caused structural damage to their property, the actual disturbance was minimal, and the work was completed with only one real setback. This was when the Fleet Ditch sewer burst and flooded the workings to a depth of ten feet as far as King's Cross.

The City of London Corporation subscribed £200,000 toward the construction of the Underground. They believed, as Pearson had intended, that it would clear the City streets of a large amount of goods' traffic. However, the Metropolitan Line was run almost exclusively as a passenger railway.

The Tube railway

THE DIFFERENCE between the Tube railway and the Underground was one of construction. The Metropolitan Underground was basically a covered ditch; the Tube was a tunnel bored beneath the surface. So, when, in 1869, the engineer Peter Barlow was given the task of boring a second Thames Tunnel (the first was Brunel's pedestrian tunnel of 1843) between Tower Hill and Bermondsey, and later a small railway was installed, this can be considered as the first Tube railway in the world. It was also the first iron-lined tunnel.

Barlow improved on Brunel's method of boring. He used a circular shield, instead of Brunel's protective box, and lined the tunnel with cast-iron segments, bolted flange to flange. The shield was driven forward by levers and jacks and averaged five feet of progress a day. The Tube had a minimum depth of 22 feet (greater than the Rotherhithe tunnel) but was only seven feet across.

Drawn by cable

It was necessary for passengers to descend by lift to the floor of the tunnel, where they took their seats in a car which was drawn through the Tube by a cable worked from a stationary engine. The car held 12 passengers.

THE FIRST CAR. *The first petrol-powered car, or rather, tricycle. Built by Karl Benz. On its first test, Benz ran straight into a wall.*

The railway was opened in August, 1870, but it was not very successful. The number of passengers it could carry was not sufficient to cover the running expenses. After only a few months the car was withdrawn and the Tube converted for pedestrian use.

The Electric Tube railway

IN VIEW OF THE FAILURE of the Tower Hill Tube rail-car to survive more than a few months, the claim to be the first *successful* Tube railway can perhaps be made by the City and South Railway, which was opened to the public on December 18th, 1890. But the City and South has another claim to fame: it was the first *electric* Tube railway in the world.

The line was originally known as the City of London and Southwark Line. It was planned to run from King William Street to the Elephant and Castle in Lambeth, a distance of one and a half miles, to be covered by cable-operated trains. But when work started in 1886, it was decided to power the trains by electricity instead. Once the limitations on the length of the railway imposed by the original need to work with a manageable length of cable no longer existed, the projected line was extended to Stockwell, a distance of three and a quarter miles.

Wooden benches

The electric locomotive weighed 12 tons and hauled the cars at an average speed of $11\frac{1}{2}$ miles an hour. These cars were extremely uncomfortable. They were designed for a tunnel that had been made as small as possible in order to reduce the costs of construction. As a result, they were confined and narrow; the passengers sat on wooden benches. The coaches were nicknamed 'padded cells' and the line became known as the 'Sardine Box Railway'. All the same, it was very popular. The fare for any distance was two pence and there were no tickets.

Towrope broke

The whole of the actual line was beneath the surface – as much as 105 feet down at the Thames crossing and never less than 45 feet down. The inclined ramp that led from the line up to the workshops and depot above ground at Stockwell was soon replaced by a lift after an unfortunate accident in which a towrope broke and a car went careering back down the slope.

All the current for the line was generated in a power house at Stockwell where three dynamos were belt-driven by steam engines.

A practical internal combustion engine

ETIENNE LENOIR, a French engineer, produced the first practical internal combustion engine in 1859. It was originally used to power factory

DAIMLER BUILDS A "MOTOR-BIKE". *A leather belt connected the engine to the rear wheel. No gears, no clutch . . . the first "bone-shaker". Daimler's real dreams, however, were of cars.*

machinery and it ran on coal gas. In 1862, Lenoir adapted the engine to run on liquid fuel and installed it in a carriage.

This strange motor completed a journey of six miles, near Paris, in three hours – considerably slower than the early steam engines of Trevithick and Stephenson! But it was the beginning of one of the most useful and yet trouble-some methods of modern transport.

Describing his experience, Lenoir wrote: *'In 1863, I made an automobile vehicle with which I often did the journey from my works through the Bois de Vincennes to Joinville-le-Pont and I took about an hour and a half to go and the same time to get back . . . The vehicle was heavy. The motor was one and a half horsepower turning about 100 r.p.m. It was fitted with a large flywheel.'*

Lenoir's engine did not last long on the roads. He very soon found out that it was better suited to propelling a boat.

The petrol-powered motor car

BY THE 1880s petrol was coming on the market from the newly-exploited American oil wells and several inventors realised its potential as a motor fuel. The first man to succeed in building a petrol-powered car was a German, Karl Benz, from Mannheim.

Benz's historic first tricycle was powered by a single cylinder, water-cooled petrol engine. The front wheel was smaller than the two back ones and could be steered by another wheel which was placed in front of the bench on which the driver and passenger sat. The use of a differential gear on the back axle made it possible for the two rear wheels of the vehicle to rotate at different speeds as the car rounded a bend.

The power of the engine was trans-mitted to the wheels by chains and the car had a very basic form of clutch which enabled the engine to run in neutral.

This was a novel form of transport. It required expert handling, which, unfortunately, Karl Benz seemed unable to provide. He ran the car straight into a wall on its first test.

Not put off by that incident, he continued to improve his invention, even though it attracted little attention at first. Recognition came in 1888 when a drive through Munich caused great excitement and was followed by a flood of orders.

The motor cycle

IN 1885, GOTTLIEB DAIMLER made the first of a series of successful high-speed internal combustion engines and installed one vertically in a crude two-wheeled 'bone-shaker' type of bicycle. This is considered to be the first real 'motor' cycle. It was exhibited for the first time at the Paris Exhibition of 1889.

The engine had an output of about half one horsepower at approximately 600 r.p.m. A small pulley on the engine crankshaft and a large grooved pulley on the rear wheel were connected by a leather belt. The drive was direct; there were no gears and no clutch mechanism.

Daimler's main interest was not with motor cycles but with cars. At one time he had been technical manager in Niklaus Otto's firm, Otto and Langden, and, in 1876, had produced the first commercially practical four-stroke gas engine. But Daimler was also experi-menting on the side with a number of petrol engines. His ideas clashed with his employers' interests in gas engines, and he was dismissed.

On his dismissal, Daimler moved to Cannstaff and throughout the autumn and winter of 1883-4 he worked on his petrol engines, intending that eventually they should be fitted to a car. His motor cycle was only an incidental stage in this research. But, by the time he had produced it, Karl Benz had already driven in a petrol-powered car.

Daimler did not produce his first car until 1886: it was four-wheeled and could achieve a speed of 11 miles an hour.

The pneumatic tyre

THE PNEUMATIC TYRE was invented in 1845 by Robert William Thomson. By increasing the comfort and safety of the passengers it was the greatest develop-ment in the wheel since its invention thousands of years before in the Middle East.

Thomson constructed an airtight inner tube made of canvas and rubber and encased in a strong cover made of a number of pieces of leather sewn together. It was reported in 1845 that a set of Thomson's tyres had been fitted to a brougham, but after that nothing more was heard of the invention.

Tyres of the metal and solid rubber type were still preferred for the next 40 or 50 years. Thomson's invention was considered too expensive and too cum-bersome to fit.

An old hose

Independently of Thomson's idea, a Belfast veterinary surgeon, John Boyd Dunlop, patented the first practical pneumatic tyre in 1887. He came across the idea quite by chance whilst making a tyre for his son's tricycle. He cut pieces from an old garden hose, glued them together and pumped them full of air.

Part 3

HOUSEHOLD

We take most of the following items for granted. Probably we do not think much about when or where they originated – the zip and the water-closet, the bra and a box of matches, the refrigerator and the carpet-sweeper – the things you come across and use almost every day.

New gadgets are constantly appearing that quickly take their place in the home as if they had always been there. Some in fact have been around for a long time.

Queen Elizabeth I used a flushing water-closet and an attempt to produce matches was made when Charles II was on the throne of England. The bra was not patented until the first year of the First World War but we have a picture of a girl in a bikini from a mosaic well over two thousands year old.

And although mammoths several times older than that have been found in glaciers perfectly frozen the first successful packaged frozen foods were not marketed until just over forty years ago.

We have, it seems, done a fairly good job in producing more efficient ways of doing things but some of our basic ideas may not always be as new as we think.

HUBERT BOOTH'S VACUUM-CLEANER. *It was enormous, a great van pulled by horses. Long hoses snaked out of the roof of the van and disappeared through first and second storey windows. Cab horses trotting by bolted in terror when Booth switched on the powerful engine hidden in the van. The whole principle was the reverse of current thinking but it worked . . . as Hubert Booth had always known it would.*

The Bra

THERE IS PLENTY OF heated argument about where the bra was first worn and who wore it. For example, there is an ancient mosaic in Sicily that shows a woman athlete wearing a bra and pants. But was *that* the first, or were there even earlier ones?

However, there is no argument at all about who took out the first patent on a bra – a one-time New York debutante called Mary Phelps Jacob, who patented the idea under the name of Caresse Crosby in November, 1914. Mary Jacob was a descendant of the steamboat pioneer Robert Fulton, and she wrote in her book, *The Passionate Years:*

'I believe my ardour for invention springs from his loins – I cannot say that the brassiere will ever take as great a place in history as the steamboat, but I did invent it . . .'

As a debutante, Mary Jacob rebelled against the box-like armour of whalebone and pink cordage that was called a corset. One night, before a ball, with the help of her French maid, she designed the original bra from two pocket handkerchiefs, some pink ribbon and some thread. Within weeks she was making bras for her friends, all immensely relieved to discover an alternative to their agonising corsets.

A stranger wrote

When a stranger wrote asking for a sample of her 'contraption', and enclosing with the request a dollar, Mary Jacob decided that the time had come to exploit her invention. She hired a designer to produce a series of drawings for 'The Backless Brassiere', and was granted a patent the following year. But the few hundred samples that she and her maid produced hardly sold at all.

Disillusioned, Mary Jacob sold her rights to the Warner Brothers Corset Company for 15,000 dollars – the patent was later valued at 15 million dollars!

Matches

THE FIRST ATTEMPTS to produce a match go back to 1680, soon after the discovery of phosphorus by Robert Boyle. His assistant, Godfrey Haukewitz, dipped pieces of wood in sulphur and struck these against phosphorus until they flared.

By the beginning of the nineteenth century, however, the usual method of striking a light was still to use a flint and steel in order to ignite tinder, which then lit a wick. It was only at that time that the idea of substituting splints of wood dipped in sulphur – as Haukewitz had demonstrated earlier – became popular. Before long the sulphur was mixed with chlorate of potash and sugar in order to improve its inflammability.

In 1830, a chemist named Jones, who had a shop in The Strand, London, offered to the public 'Promethean matches' made from a roll of paper which had the inflammable mixture at one end with a small, hermetically sealed tube containing a minute quantity of sulphuric acid. A pair of pliers was also supplied, with which the purchaser was to crush the end of the tube: the acid then came into contact with the mixture and ignited. It sounds – and probably was – tricky and dangerous.

Friction matches

Three years later, the first friction matches appeared. They were known as 'Lucifers' and the name stuck for nearly a century. These matches were coated with sulphide of antimony and chloride of potash made into a paste with gum water. They were lit by being drawn between the two surfaces of a folded piece of sandpaper.

The final step was to substitute phosphorus for antimony – then the 'Congreve' match was born, named after a famous rocket expert. This sold at a shilling a box, and did a good trade in Britain and overseas.

Safety matches were introduced in Sweden in 1852. They made use of the less dangerous amorphous phosphorus.

The Zip-fastener

'C-CURITY' WAS THE BRAND NAME given to their product by The Automatic Hook & Eye Company of Hoboken, New Jersey. But the zips were not secure at all. Due to their habit of springing open at awkward moments, they had to be sold from door to door: the garment manufacturers would not touch them.

'C-curity' were not the first zips, however. The inventor of the zip was Whitcomb Judson, an American mechanical engineer, who filled the first patent applications in 1891 and 1894. His fastener was originally designed for shoes, and consisted of a series of separate fasteners each having two interlocking parts that could be fastened either by hand or with a movable guide.

This idea caught the interest of an acquaintance of Judson, a corporation lawyer called Colonel Lewis Walker. Colonel Walker set up The Universal Fastener Company of New York to make the Judson fastener. At first the zips were made by hand; later, money was raised to develop a suitable machine.

A new fastener

By 1905, Judson had designed a new fastener, easier to produce on a machine. The fastening elements were clamped to the edge of the fabric tape instead of being linked together in a chain. It was then that The Universal was re-organised as The Automatic Hook & Eye Company.

In a determined effort to break into the market, the company engaged Gideon Sundback, who not only improved the fastener but also designed more efficient machines to make it. Even so, the first big contract was not until 1918, when a contractor making flying suits for the U.S. Navy ordered 10,000 fasteners. Five years later, B. F. Goodrich put the fasteners into the galoshes for which they are famous. From then on, the popularity of the zip was assured.

Margarine

THE 'BUTTER-SUBSTITUTE' was patented in 1869 by Hippolyte Mège Mouriés, who had been commissioned by the French Navy to find an alternative to butter at a time of acute shortage. Hippolyte was handsomely rewarded for his achievement by the French King, Napoleon III.

In his experiments, Hippolyte tried to imitate what he thought were the actions of a cow's udder – the conversion of body fat into the fat of the milk. To do this, he crushed beef suet with sheep's stomach at first and later with cow's udder and added warm milk; finally he pressed out the fatty content between warm plates! The result had a fairly low boiling point.

He called his product 'Margarine' after the Greek word '*margaron*', meaning 'a pearl', because of the pearl-like appearance of his original preparation.

Manufacturers of Margarine soon found that they could do without adding the cow's udder. The Margarine still proved so successful with the public that the London *Times* of September 4th, 1873, predicted that it would soon drive butter off the market.

"PROMETHEAN" MATCHES. *Containing sulphuric acid, these matches were introduced in 1830.*

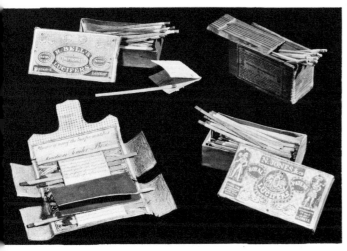

"LUCIFERS". *The first friction matches. They were used from 1833 into the twentieth century.*

ABOVE. "CONGREVES". *The first modern phosphorus matches were named after a famous rocket expert.*

RIGHT. THE FIRST BRA ? *This Ancient Roman mosaic is the oldest known representation of a bra.*

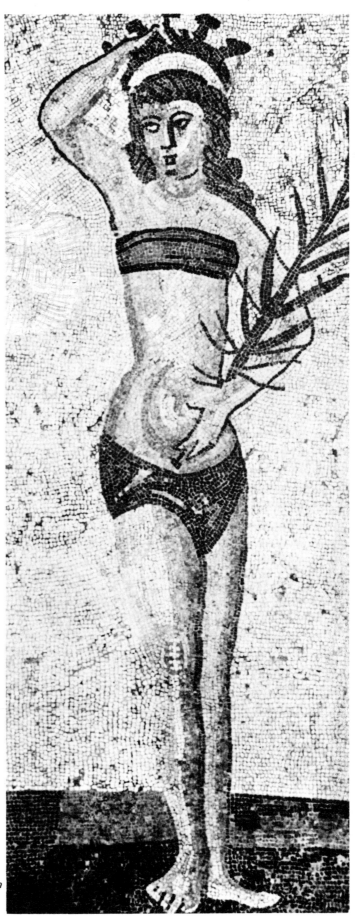

The water-closet

WATER-CLOSETS OF ONE sort or another have been used for thousands of years: for instance, seat-closets dating from 2000 B.C. have been unearthed in Mesopotamia; these were built of stone and connected by a sloping conduit to a street drain.

Closets built over rivers or streams were used by the inhabitants of Knossos, in Crete, 1400 years before Christ, and used also by the Greeks and later the Romans.

Medieval Europe was far less sanitary. In the fourteenth and fifteenth centuries, London had only a dozen public latrines, some of them built over the Thames. However, in those days, most privies had to be emptied by servants and collected in the early morning by 'the night soil men', who took away the contents in carts and tipped it into cesspools.

Punning title

In Elizabethan times, a godson of the Queen, Sir John Harington, first thought of an idea for an odourless closet, which he described in a book he published in 1596. The book had the punning title, *Metamorphos is of Ajax* (a 'jakes' was a popular name for a privy).

Harington's closet had a bowl which could be filled with water from a cistern to a covering depth of two feet, and which could be emptied through an underlying valve into the cesspit. He tried – without much success – to seal off the cesspit from above except for the moment of emptying. Plenty of smell remained.

Nonetheless, Elizabeth I was impressed and installed one of the first valve water-closets at Richmond.

Nearly 200 years later, in 1775, Alexander Cumming solved the problem of sealing off the cesspit. The principle underlying his 'Water Closet upon a new Construction' has been in use ever since: the soil pipe immediately below the pan was bent into an 's' shape 'so as constantly to retain a quantity of water to cut off all communication of smell from below'.

300 patents

Cumming's invention started a wave of enthusiasm: in the next 100 years, almost 300 patents were taken out on various types of water-closets.

One invention was particularly important. It was necessary to control the supply of water to the pan automatically. A major obstacle to the more widespread

MACHINE-MADE CLOTHING. *Thimmonier at his sewing machine.*

use of the water-closet had been the objection of the water supply companies to the extravagant waste of water. This problem was solved by Rogers Field, who designed a Syphon Automatic Flushing Tank.

The sewing machine

PATENTS FOR THE EARLIEST sewing machine were taken out in 1790 by Thomas Saint, a London cabinet-maker, and remained unnoticed for 84 years. Saint's invention was a chain-stitch machine which had a forked needle following behind an awl (the small tool for pricking the cloth). According to his patents, the machine was intended for '*stitching, quilting or sewing to be worked by hand, a Mill,*

Steam Engine or other power'.

The first chain-stitch machine to have any real success was made in 1830 by French tailor, Barthelemy Thimmonier, not a name to be forgotten easily. It was cumbersome and made mostly of wood and its barbed needle tended to tear the fabric. Nonetheless Thimmonier was confident enough in his invention to set up a clothing factory in Paris, in which he put 80 of his machines to work for the French Army. Unfortunately the workshop was wrecked and the machines were broken up by manual clothing workers who feared that they would lose their jobs to the new devices.

Thimmonier was not to be put off. He improved the design of his machine and in 1848, he took out patents in England and America. All the same, he died without any money, even though he was responsible for a considerable revolution in the manufacture of clothing.

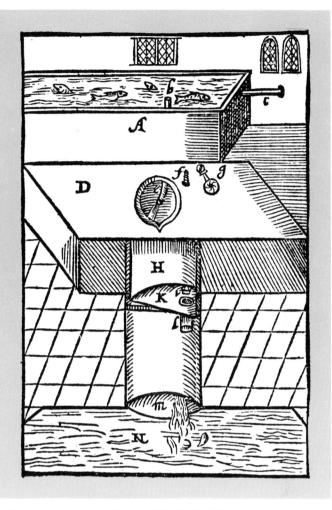

SIR JOHN'S WATER CLOSET. *In 1596 Sir John Harington published a book outlining an odourless privy. Asked by the Devil how he dare to pray on the W.C., the wise man answers that the prayers are for Heaven, but the filth is safely bound for Hell. This water closet worked with the aid of a cistern which washed the bowl's contents down to a cess pit.*

Refrigeration

REFRIGERATION, AS WE KNOW IT today in the form of the domestic 'refrigerator', was first achieved one summer evening in the 1830s by Jacob Perkins, an American who spent most of his life in England.

On that evening his working model using what was known as a 'Vapour Compression Cycle') succeeded in making a small amount of ice, which Perkins's excited mechanic wrapped in a blanket and took by taxi across London to his master's lodgings.

Perkins, however, never developed his invention. But shortly after, on the other side of the world and working along the same lines, James Harrison, a Scottish printer who had emigrated to Australia, first noticed the cooling effect of ether while using it to wash down type faces. Using this discovery he designed a machine that was made by Daniel Siebe and shown at the International Exhibition of 1862. Machines to that design were the first refrigerators to be put on the market.

Meatships

Harrison's experiments had been urged on by the need to get Australian meat to Britain still fresh, but his first attempts to ship it over failed. However by the turn of the century refrigeration was being used regularly to preserve food and for industrial purposes.

Later an American, Willis H. Carrier, demonstrated how it could be used to provide comfortable conditions inside buildings.

The deep freeze

IT IS SAID THAT SIR FRANCIS BACON died from a chill which he caught when trying to freeze a chicken by stuffing it with snow. At any rate, long before then – since the very earliest ages – men have realised the value of cold for preserving food.

Not until after the First World War, however, was it confirmed that the key to successful preservation lay more in the speed of freezing than in achieving extremes of cold.

After a visit to Labrador, Clarence Birdseye made experiments on rabbit meat in his own kitchen, then in a refrigeration plant in New Jersey. He developed a process in which cartons of food were pressed between refrigerated plates.

The principle behind this method is still in use, although the plates have been replaced by 'blast freezing' in a wind tunnel.

The first retail sales of packaged frozen foods were at Springfield, Massachusetts, in 1930. Birdseye died 26 years later, at the age of 70, with 300 patents to his credit.

The carpet-sweeper

ATTEMPTS AT MECHANICAL sweeping probably begun in 1811 when an English patent was issued to James Hume for a sweeper. But nothing much came of Hume's machine.

The first practical carpet-sweeper to appear on the general market was invented by the owner of a china shop in Grand Rapids, Michigan, who suffered from headaches brought on – so he thought – by the dusty straw in which the china was packed.

To cure his headaches, M. R. Bissell designed a sweeper in 1876 with a box to hold the dust and a knob to adjust the brushes to variations in the surface of the floor.

Bissell and his wife demonstrated their sweeper at church socials – with great success. Most of the components were made by women working at home and then assembled by Mr and Mrs Bissell.

The public were ready and anxious for just such a machine: Louis Pasteur's theories on germs and their dangerous effects had recently made women very hygiene-conscious. In fact the new sweeper became so popular that people spoke of 'bisselling' the carpet in the same way that many today might talk of 'hoovering'.

The vacuum cleaner

A NUMBER OF INVENTORS at the end of the last century had become very interested in the possibilities of cleaning carpets and the inside of houses either by blowing air at the carpets or by sucking air in – and hopefully the dust as well.

A demonstration was carried out in 1901 at St Pancras Station, London, of a new cleaner designed for use in railway carriages. This cleaner was based on the principle of blowing dust out: its effect was not very successful. One of those who attended the demonstration, Hubert Booth, suggested that sucking the dust in might be better, but he was told firmly – and possibly with some indignation – that it had been tried and did not work.

He sucked hard

Back at his home, Booth lay on the floor with a handkerchief over his mouth and sucked hard. The dirt trapped in his handkerchief reassured him that he was right: what was needed was only an efficient cloth filter that would trap the

dust and let the air through.

His first vacuum-cleaner was a monstrous contraption drawn by horses, It stood in the road and sucked dirt from inside the house through a long hose. Not surprisingly, Booth had trouble with the police: the 'noisy serpent', as it was called, caused passing cab-horses to bolt in terror.

To this day, all cylinder vacuum cleaners rely on the principle used by Booth.

The safety razor

IT WAS WILLIAM HENSON, of London, who first designed the safety razor in the modern form. In 1847 he took out a patent for what he called 'Certain Improvements in the Construction of Razors for Shaving'.

According to the specification of the instrument, Henson's 'comb tooth guard or protector' was attached to an ordinary straight razor or to a razor 'the cutting blade of which is at right angles with the handle and resembles somewhat the form of the common hoe'. There was one major drawback to the razor: a small hollow ground blade was used, which had to be fixed in a separate holder. Henson himself did not use the term 'safety razor'; according to the Oxford English Dictionary, the term did not appear in print for another 46 years.

Flint blades

As for earlier razors, we cannot tell when man first began to use a knife or blade to shave the hair from his face. Blades made from flint have been found at sites in Egypt dating from thousands of years before Christ, and archaeologists have identified these blades as razors.

The first *attempt* at a safety razor seems to have been made in Paris, in the Eighteenth Century, by Jean-Jaques Perret, a master-cutler, who got fed up with contracting skin-diseases after 'false-shaves' at his barber's. In 1762, he added a wooden guard to a straight razor, and illustrated this in a book called, *L'art du Coutelier*. However, the razor was far too expensive for general use.

Throw-away razor blades

HENSON'S SAFETY RAZORS were all very well, but they needed to be sharpened and

that was a long and tedious task requiring quite a lot of skill. Something easier and quicker was needed.

The man who supplied the solution was King C. Gillette, of Boston, Massachusetts. He proposed using mass-produced, wafer-thin blades that would be so cheap that they could be thrown away as soon as they became blunt. The idea of the wafer-thin blade was not new. As early as 1814 – before Henson's razor – the firm of Rhodes and Champion, of Sheffield, had been making a *non*-safety razor which consisted of just such a blade held in a frame.

Gillette's razor, which immediately became immensely popular, was first brought out in 1895.

Twentieth century wafer blades are made from continuous strips of steel, each weighing 30 pounds and long enough for 12,000 blades. The strips are passed through a series of machines, during which they are perforated, toughened, blued and tempered, then lacquered and passed through yet another machine which produces the cutting edge and severs each blade.

Food preserves

THE FOLLOWING LETTER marked the highest approval for Bryan Donkin's canned food, which he daringly sent for tasting to Britain's Commander-in-Chief the Duke of York. In June, 1813, Bryan Donkin and his partners received this note from Kensington Palace:

'Gentlemen, I am commanded by the Duke of Kent to aquaint you that His Royal Highness, having procured introduction of some of your patent beef on the Duke of York's table, where it was tasted by the Queen, the Prince-Regent and several distinguished personages and highly approved, he wishes you to furnish him with some of your printed papers in order that His Majesty and many other individuals may, according to their wish expressed, have an opportunity of further proving the merits of the things for general adoption.

'Your most obedient servant, Jon Parker.'

Donkin, however, owed a good deal of his success to Nicolas-Francois Appert, a pastry-cook from Paris. Appert had considered the particular need for preserved foods on board ships and had formed some pet theories on fermentation. As a result, he devised a bottling system which

Because Mr Bissell had a headache . . .

he put into practice in 1806, in a small factory on the outskirts of Paris.

Trial at sea

Appert's method was to pack his foods in special wide-mouthed bottles which were tightly corked and sealed and then sterilised by being placed in a bath of water which was then brought to the boil. He experimented with meat, vegetables, eggs, fruit, fruit juices and even mushrooms.

On December 2nd, in the same year, French ships took aboard 18 different foods which had been preserved in this way by Appert. The ships remained at sea until April 13th, the following year, when the containers were then opened: all the food was found to be in good condition and perfectly edible.

The French naval authorities were most impressed and offered Appert 12,000 francs on condition that he explained his methods in print. Appert wrote *The Art of Conserving for Years All Animal and Vegetable Substances* in 1810; it was later published in London. But Appert must have kept some of his secrets back, because the book caught the eye of Bryan Donkin, an English engineer and manufacturer, and his two partners, Hall and Gamble, and these three thought it worthwhile to pay Appert the sum of £1,000 for further details of his methods.

Donkin's idea

Donkin realised immediately that boiling the bottles in a bath of water was not a very reliable way of sterilisation. Experiments in his engineering factory led him to raise the temperature of the bath still higher by adding chloride of lime to the water. This ensured complete sterilisation.

But Donkin was still not satisfied. The glass containers were clearly too fragile; so he replaced them by cans of tinned iron. These tin-plated cans were hand-made and heavy. They were filled through a hole in one end over which a tinned iron disc was then soldered – a hammer and chisel were needed to open the cans!

The can-opener

The chisel-and-hammer method of opening lasted for 50 years, until thinner steel, with a rim round the top of the can, meant that more practical can-openers could be used. The earliest known domestic opener was the bull's head type used on cans of bully beef. The first appearance of a can-opener in the catalogue of the Army and Navy Stores was in 1885.

The vacuum flask

THE ORIGINAL IDEA for a vacuum flask can be traced back to suggestions made by an English inventor, Lewis Gompertz, in a book he wrote in 1850, called, 'Mechanical Inventions and Suggestions'. Among the suggestions were some 'To Produce a Fire-proof Box', which outlined the fundamental idea of the vacuum flask. There is no evidence, however, that Gompertz ever tried to make his box.

The vacuum flask itself was designed in Germany, about 50 years later, in 1904, by Reinhold Burger. Burger's design was in fact based on experiments made by the British professor James Dewar some years earlier. In order to find a suitable name for Burger's flask, the Germans held a competition, which was won by the man who suggested the Greek word for 'heat' – 'thermos'.

Dewar's discovery

Although James Dewar had no share in marketing the idea, he had worked out the principles which Burger adopted. It was Dewar who found he could reduce the transfer of heat from a container by surrounding the container with a high vacuum, and it was he who also discovered that he could further reduce the transfer of heat by silvering the inner walls of the container. Finally it was Dewar who improved his own idea by adding a small quantity of charcoal in the space which separated the walls of the container.

The pressure-cooker

THE PRESSURE-COOKER is not so recent an invention as one might think. The 'New Digester or Engine for Softening Bones' was first demonstrated in 1680 by a Frenchman, Denis Papin; and on April 12th, 1682, members of the Royal Society sat down to a meal cooked in the 'Digester'.

John Evelyn, the great diarist of that time, wrote:

'*I went this afternoon with severall of the Royal Society to a supper which was all dressed, both fish and flesh, in Monsieur Papin's Digesters, by which the hardest bones of beefe itselfe and mutton were made as soft as cheese, without water or other liquor, and with less than eight ounces of coales, producing an incredible quantity of gravy; and for the close of all a jelly made of the bones of beefe, the best for clearness and good relish, and the most delicious that I have ever seene or tasted.*

'*We eat pike and other fish bones, and all without impediment; but nothing exceeded the pigeons which tasted just as if baked in a pie, all these being stewed in their own juice, without any addition of water save what swam about the Digester . . .*'

The principle behind the pressure-cooker is that the temperature of steam rises under pressure. In the modern pressure-cooker, the pressure is usually kept to a maximum of 15 pounds per square inch. Papin's Digester was a good deal more powerful: bones softened at 35 pounds per square inch and almost disintegrated at 50 pounds per square inch. It was made of cast-iron

Safety-valve

Papin's boiler was in fact equipped with a steam safety-valve controlled by sliding a weight along a lever, but such cookers were dangerous because of crude engineering techniques Pressure-cookers only became manageable when they were made of light aluminium: now, of course, they have much more accurate means of control, their being made of cast-iron did not make them any less dangerous.

Micro-wave cooking

THIS IS THE FIRST COMPLETELY new method of cooking since prehistoric man discovered how to make fire.

In this revolutionary technique there is no direct or indirect application of fire to food. Instead, the food is bombarded with electro-magnetic waves; this causes molecular activity, which produces the heat for cooking. It may sound complicated but it is extremely effective.

Link with radar

The original idea for the cooker came from experiments linked with Britain's radar defences during the Second World War, and the electronic tube that supplies the micro-wave energy was invented by Dr. H. A. H. Boot and developed at Birmingham University. The first micro-wave cooker was produced in the late 1940s by the American firm of Raytheon Incorporated.

THE SEA

*All the major voyages of discovery, the first circumnavigations
of the globe and the great migrations of the South Sea islanders
across the Pacific, were undertaken by sail.*

*Sails and oars were the only means of travel at sea from the
time of the Ancient Egyptians to Cleopatra's barge, from the
Greeks at Salamis to the Roman slave galleys, from the Viking
longboats to the Barbary pirates, from the Kon Tiki to Columbus,
from the 'wooden walls' of Elizabethan England to the* Mayflower,
and from the Bounty *to the* Cutty Sark.

*Steam power brought an end to this long tradition. Once again,
just as on the land, the power of steam transformed travel and
transport. Ships became more powerful, they went faster, they
grew larger and larger: the paddle steamer, the propeller-screw,
the hovercraft and nuclear power – each in turn a spectacular
advance that reduced the vast expanses of the ocean almost
to a lake.*

THE FIRST REGULAR OCEAN STEAMER. *The "Great Western" was specially built for transatlantic service. She was made of wood
– with the hull sheathed in copper below the water-line.*

The steamboat

THE *Charlotte Dundas* made five knots on her maiden voyage. She was built in Scotland, in 1801, and she was the first *practical* steamboat, though not the first known vessel to be driven by steam.

A Frenchman, Jaques Perier, won that honour for the boat he tested on the river Seine, near Paris, 26 years earlier. But the steam engines of Perier's time were not nearly powerful enough to drive their own tremendous weight, and Perier's trials were not a success.

William Symington designed and built the engine for the *Charlotte Dundas* and named her after the daughter of his patron, Lord Dundas. He had begun work on an experimental boat several years before, with the encouragement of Patrick Miller, an Edinburgh banker, but Miller had lost interest and Symington had been left to press on with his designs alone, until the Dundas family came to his help.

Canal damaged

The *Charlotte Dundas* was used to haul barges along the Forth and Clyde Canal. She was 56 feet long and 18 feet in the beam. The engine was fixed on the port side and the boiler which supplied the steam to work the engine was on the starboard side. There was only one paddle wheel, as you can see, at the stern.

Unfortunately her career as a tug was brought to an end when it was discovered that her wash damaged the canal banks, a problem that we still have today when the wash from cruise boats steadily wears away the river banks. She was broken up at the age of sixty.

A steamboat passenger service

THE SINGLE FUNNEL of the *Comet* was also used as a mast. It carried a yard and a square sail. With this added power, the *Comet* could steam along at six or seven knots.

She was the first steamer in Europe to run a commercial passenger service. The service ran along the river Clyde from Glasgow to Greenock. A first class cabin cost four shillings, but the excitement of the novelty was probably worth a great deal more.

The Comet was built at Glasgow by Henry Bell and began service in August,

1812. She was built of wood, and after a lot of early trouble, had two paddle wheels fitted. Her breadth over the paddle boxes was 15 feet and she was 51 feet long – smaller than the *Charlotte Dundas*.

As with the *Charlotte Dundas,* her engine was set on the port side and the boiler on the starboard. In order to make the boiler more stable, it was fixed into a brickwork frame.

Six years after coming into service the *Comet* also began steaming between Glasgow and the Western Highlands. Two years later, however, she ran ashore at Craignish Point and became a total wreck.

An Atlantic crossing by steamboat

THE FIRST STEAM-PROPELLED vessel to cross the Atlantic was the American ship *Savannah*. But the *Savannah* was not a proper steamboat. She was a fully-rigged sailing boat which had been fitted with an engine and detachable paddle wheels. These could be stowed on deck when they were not in use.

The *Savannah* left her home port New York, on May 24th, 1819 – carrying no passengers – and docked at Liverpool on June 20th, having taken 27 days and 11 hours for the crossing. From Liverpool

ABOVE: STEAMING UP THE RIVER ... *the "Charlotte Dundas", the world's first practical steamboat. She was 56 feet long. Her single paddle wheel was lodged under the deck, at the stern.*

RIGHT: PLAN OF THE "CHARLOTTE DUNDAS" *Scottish engineer, William Symington, designed and built the "Charlotte Dundas" in 1801. The engine was on the port side and the boiler, starboard.*

ne went to Russia, where it was hoped
 sell her to Tsar Alexander I. He made
trip in her, but refused to buy her.
   Two years later, the *Savannah* was
 recked on the shore of Long Island,
 Jew York.

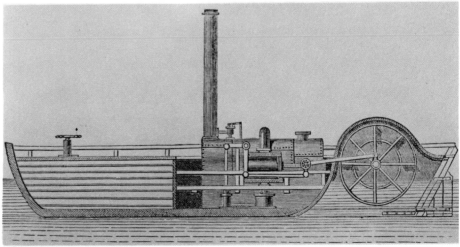

NEW YORK WELCOMES THE "SIRIUS". *Built in 1837, the "Sirius" was the first ship to cross an ocean under steam. Carrying 40 passengers, she reached New York on 22 April, 1838. The trip took 18½ days. Average speed was 6.7 knots.*

An Atlantic crossing under continuous steam

THE FIRST STEAMER to cross the Atlantic under continuous steam and so perhaps earn for herself the title of the first real ocean 'steamer' was still built of wood, surprising as that may seem.

The *Sirius* was built in 1837, at Leith, near Edinburgh. She was chartered by the British and American Steam Navigation Company, and left Cork Harbour with 40 passengers on April 4th, the following year. She reached New York in 18 days and 10 hours, at an average of 6.7 knots – a good speed.

Fresh water

The success of the *Sirius* had depended on an invention by Samuel Hall, three years before she was built. Previous boats that had attempted to cross the Atlantic under steam had had to feed their boilers with sea water and had not been able to carry enough fuel to convert this to steam.

Samuel Hall's invention – the 'surface condenser' – enabled marine boilers to b fed with fresh water that was recircuite so that the boilers could be kept i continuous operation.

An iron paddle steamer

THE FIRST PADDLE STEAMERS had bee made of wood. But timber for ship building was becoming scarce and expen sive. As a result, and for reasons o

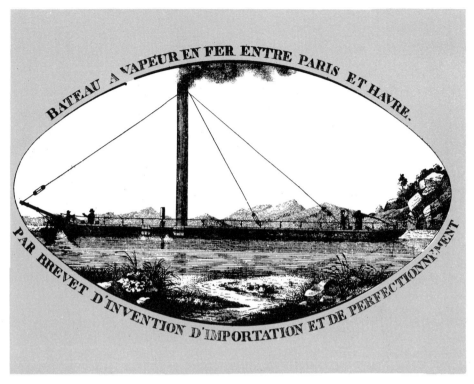

THE IRONCLADS. *1821 saw the birth of the first iron paddle steamer, the "Aaron Manby". The ironclads soon replaced the wood-built ships. Thirty-seven years later, David Livingstone sailed up the Zambesi River in the first steel paddle-launch.*

strength as well, naval architects began experimenting with iron.

The first iron paddle steamer was the *Aaron Manby,* built in 1821. 'Ironclads', as they were called, very quickly replaced wooden ships completely.

Thirty-seven years later the first steel paddle launch – the *Ma Robert,* built by Laird Bros. – was used by David Livingstone for his expedition up the river Zambesi, in Africa.

The steel hull weighed 13 tons and had a 12 horsepower engine. The engine was obviously noisy and probably not very smooth, because the *Ma Robert* earned the nickname 'Asthmatic'.

A Transatlantic service

THE *Great Western* was the first steamer specially to be built for transatlantic service. She was built by the famous engineer Isambard Kingdom Brunel and launched at Bristol on July 19th, 1837, for the Great Western Steamship Company. She completed her maiden voyage across the Atlantic in April, the following year, shortly after the crossing of the *Sirius.*

The *Great Western* was made mostly of wood. Her ribs were of oak and bolted together in pairs. There were also four rows of iron bolts that ran lengthways through the bottom frames of the ship. As further protection, the hull was sheathed in copper below the waterline.

A screw-propelled ironclad

THE *Great Britain* took almost three years to build. Like the *Great Western,* she was designed by Isambard Brunel for the Great Western Company, and she was completed six years after the *Great Western.*

The *Great Britain* was the first iron-built, screw-propelled vessel to cross the Atlantic under steam. Once iron had come

into use and the efficiency of engines had been improved, steamships increased greatly in size. It had already been realised that paddle-wheels were too frail and easy a target in war; now it was seen that they were too clumsy for the big new ships.

The propeller was the answer to both problems. By the time the *Great Britain* was built, several boats had already been fitted with propellers, but none had yet dared to cross the Atlantic.

Cruel fate

The hull of the *Great Britain* was 322 feet long and divided into six water-tight compartments, in case she sprang a leak. There was room for 360 passengers in all. A six-bladed propeller was fitted and she reached a speed of 11 knots on trials.

She left Liverpool on July 26th, 1845, carrying only 60 passengers but 600 tons of cargo, and arrived in New York after a crossing of 14 days and 21 hours at an average of 9.3 knots.

Seven years later, the *Great Britain* entered the Australian trade, which she plied for 20 years. A cruel fate awaited her in the end. Her engines were removed in 1882 and she was converted into a sailing ship.

The steam turbine

THE STEAM TURBINE – turning a wheel by means of jets of steam – is not a recent invention. Hero of Alexandria thought of it 1800 years ago!

Hero built a tiny boiler from which steam squirted out of two bent-armed jets mounted on a pivot, rather like the lawn sprinkler of today. We call this a 'reaction' turbine: the wheel was forced round by the steam shooting out in the opposite direction. Hero called it an 'aeolipile', after Aeolus, the Greek god of wind.

Impact method

Hero never really applied his invention to any use, and it was not until the end of the nineteenth century that a Swedish engineer, Gustav de Laval, built turbines for driving small machines. Even so, Laval's steam turbines were a little different to Hero's. Laval's worked by aiming a jet of steam directly *at* a wheel in order to turn it – what we call an 'impulse' or 'impact' turbine.

The first turbine-powered ship was designed by the Hon. Charles Parsons, the son of one of Britain's most famous astronomers, in 1885. It was called the *Turbinia,* and was 100 feet long and powered by a 2,000 horsepower engine.

The warning

The *Turbinia* gave an unofficial demonstration of its prowess at Queen Victoria's Diamond Jubilee celebrations when it 'gate-crashed' a naval review at Spithead. A picket boat was ordered to warn her away, but to the embarrassment of the Admiralty the picket was unable to catch up with the *Turbinia.*

The incident provided plenty of publicity and, perhaps in order not to be caught out again, the Admiralty themselves ordered two turbine-driven destroyers.

The chronometer

IN 1714 the British Government passed an Act of Parliament offering £10,000 to anyone who should discover a method of determining longitude within one degree or 60 miles, £15,000 if the method proved accurate to within 40 miles, and £20,000 if it proved accurate to within 30 miles.

Of all the inventors who competed for the prize – and they included Sir Christopher Wren – it was a carpenter, John Harrison, who succeeded. It took him seven years to make the world's first chronometer – and 45 years to get his full reward.

The chronometer is a clock that measures time with great accuracy. It is used for navigation at sea, and gives the time for observations of the stars and planets for determining the position of a ship. It is not affected by changes in temperature and is balanced so that it remains horizontal however much the ship rolls to one side or the other or lifts either at the bow or the stern. Even the best ordinary watches are usually affected by temperature changes and violent changes of position.

Extra dials

The chronometer is set at Greenwich mean time. Its face is the same as an ordinary watch or clock, with a minute and an hour hand, but it has two extra dials – one with a second hand which moves every half second, the other with a hand which shows the number of hours since the last winding.

Harrison submitted his chronometer to the Royal Society in London in 1735 and was given a certificate stating that the principles of his invention promised a great degree of exactness. The chronometer was given a successful trial on board the warship *Centurion.*

£20,000 prize

Armed with a letter from the Navy praising the accurateness of the invention, Harrison then went before the 'Board of Longitude' expecting to receive the £20,000 prize. Instead he was voted £500 expenses toward the cost of designing a less cumbersome instrument. It was not until he petitioned Parliament that an Act was passed awarding him £5,000. Even so, the Board still haggled and paid him only a very small part of this sum.

When he was awarded the Royal Society's Gold Medal, he got another £1,000 of the prize money out of the reluctant Board, and a year later they promised to give him the full £20,000. But it was not until 1773 that he received the last of the money. He was then 80, and he died three years later.

Captain Cook tested one of Harrison's smaller chronometers on his voyage round the world.

The lifeboat

THE INVENTION of the lifeboat is often linked with the name of Henry Greathead, but this seems to be a little unfair on the work of one or two other people.

The origin and the development of the lifeboat is also said to be 'as English as the cliffs of Dover', but the earliest recorded experiments for an unsinkable boat were in fact made by M. de Bernières, Comptroller-General of Roads and Bridges in France, who invented a boat that would not sink or capsize when filled with water. That was in 1765, but we have no evidence that his boat was ever put to practical use.

The first Englishman to attempt to make an 'unimmergible' boat was Lionel Lukin, a London coach-builder, who was involved with the first known attempt to establish a lifeboat station on English shores. In 1786, a 'coble' – a kind of fishing boat common in the north – was sent to him from Northumberland to be converted into an 'unimmergible' boat and was later used for saving shipwrecked sailors.

The well

At about the same time, several other people were interested in building an unsinkable boat. Henry Greathead was one of these. Another was William Wouldhave, a housepainter and singing teacher from South Shields, who had been struggling with a design for some years. Unlike M. de Bernières and Lionel Lukin and possibly Henry Greathead, however, Wouldhave wanted to make a boat specifically for rescue work.

His idea for the right design came one day in 1789 when he was watching a woman drawing water from a well. The woman had a small wooden dish which she dropped into her bucket when it was full. Try as he might, Wouldhave could not turn the dish upside down; every time, it bobbed upright again.

This was a lucky encounter. With the shape of the dish in mind, Wouldhave made a model with a straight keel and high, peaked ends fitted with water-tight cases containing cork.

Reward offered

Shortly after this a group of South Shields citizens, calling themselves 'The Gentlemen of the Lawe House' offered a reward in the *Newcastle Courant* for a model of a boat that could keep afloat in stormy weather. Both William Would-

THE FIRST "UNSINKABLE". *Well, there were several, all around the same time, but this is certainly the first regular lifeboat –
the "Original". It was launched in 1790.*

have and Henry Greathead submitted models and plans. The 'Gentlemen' did not approve either of them but, as consolation, they offered half the prize of £2 to Wouldhave. Greatly insulted, he marched out, leaving his model behind.

Model in clay

Using his design, the 'Gentlemen' then made a model of the boat in clay and invited Greathead to build the boat, which he did, giving it a curved keel – his only contribution to the design.

This boat was completed before the end of the year. It was named the *Original* and was launched in January, 1790 – the first-ever regular lifeboat. During 40 years of service, the *Original* saved many hundreds of lives, but she was dashed to pieces finally on the rocks in 1830.

The hovercraft

THE HOVERCRAFT began with a cat-food tin, a coffee can and a vacuum cleaner.

Christopher Sydney Cockerell, an electronics engineer and part-time boat builder, had a theory that ships could travel much faster if there was no wave resistance. He had noticed, for instance, that the steel runners fitted to skates and sleighs not only created friction as they slid across the snow but at the same time melted a thin layer of ice to form a film of water that served as a lubricant for the runners. He decided to apply this idea to ships.

'In designing a new boat', he wrote, 'wave resistance and skin resistance are alternative evils. To reduce wave resistance one must increase the area of the boat and therefore its skin resistance –

and vice-versa. If I could make the skin of my craft a film of air between hull and water, the skin friction would be negligible and I would then be free to design entirely around the problem of wave resistance'.

So he fitted a cat-food tin inside a coffee can and used a vacuum cleaner to blow (not suck) air down the ring between them and out at the bottom.

Jet of air

When the vacuum cleaner's jet by itself was aimed directly on a pair of scales, the pressure was one pound. But the pressure of the jet of air that came out of the bottom of the home-made gadget was three pounds at least. This increase in pressure by squeezing the air through a narrow outlet meant that a fairly heavy craft could probably be supported clear

of the ground on a cushion of pressurised air. That was the principle on which the hovercraft worked.

Cockerell tried to 'sell' his hovercraft idea without much success between 1953 and 1956. A neighbouring boat-builder helped him construct a balsa-wood model which was first tested on a lawn and later on a lake. Finally a demonstration was arranged for the Ministry of Supply, the project was classified as being suitable for military application and a contract was placed with Saunders-Roe.

The first hovercraft, SRN1, was built in eight months and had its maiden flight across the English Channel on July 25th, 1959.

Narrow slots

The hovercraft is rather like a squashed trumpet. On top there is a large funnel – or several funnels – which spreads out below to an exaggerated mouth. The air is drawn in at the top and pushed out at the bottom, where it acts as an air-cushion to support the craft. What happens in between is important: when the air is sucked in it is slowed down so that it comes out *under pressure* (in the SR-N1 this is 15 pounds per square foot). To increase the pressure the air is only allowed out through narrow slots round the edge of the craft's flat bottom.

The angle at which the air is squirted out is also important. By aiming it inwards from the edge, Cockerell made sure that the air produced the thickest possible cushion for the power that was available.

One great improvement was made to the SR-N1 the following year. It was fitted with a flexible skirt which sealed in the air cushion and conserved power; it also increased obstacle clearance.

Nuclear-power ships

THE FIRST SHIP to be driven by nuclear power was the United States submarine *Nautilus*. During sea trials in 1955 she sailed 60,000 miles on a single charge of fuel.

The first nuclear-powered surface ship to take to sea was the Soviet ice-breaker *Lenin*, in 1959. Her engines produce 50,000 horsepower, and she operates throughout the season to keep the northern sea-lanes open for shipping. She is 440 feet long and has a top speed of 18 knots.

The world's first nuclear-powered merchant ship was the *Savannah*, which was built in New Jersey with accommodation for up to 60 passengers and 10,000 tons of cargo. She ran successful trials in April, 1962. The United States Government subsidised the construction of the *Savannah* in order to show how nuclear power could be applied to peaceful use.

Cooled by water

In a nuclear-powered ship heat is generated in an atomic pile and made available for propulsion in a number of ways. The pile is either cooled by water circulated under very high pressure or by carbon dioxide gas. In each case an external heat-exchanger is used for raising steam and from then on the plant is of the conventional steam pattern.

At present, nuclear power is not very economical. But in matters of defence – such as submarines – it is ideal because the craft can develop full power under water for an indefinite time, limited only by the amount of oxygen needed on board for the crew.

ABOVE: THE BEGINNING OF HOVERCRAFT. *This is how Christopher Cockerell developed his first ideas for hovercraft—with a cat-food tin, a coffee tin and a vacuum cleaner.*

LEFT: THE UNITED STATES BUILT IT... *the "Nautilus", the world's first nuclear-powered submarine. During a 1955 test, she sailed 60,000 miles without re-fueling.*

Part 5
MEDICINE

There was a time, not so long ago, when the dentist went to work with pliers and wire – no drill, no injections, nothing to kill the pain: if the tooth was bad, out it came! A visit to the dentist was a lot more terrifying than we can perhaps realise.

A visit to the surgeon was likely to be worse. If a wound was much more than a cut, it would probably fester; the limb would have to be amputated – with a knife, without anaesthetics. There were thousands of limbless soldiers, bandaged and out of work, continuously begging their way across Europe, from the Middle Ages to the Napoleonic wars.

Or there was the plague, uncontrollable and incurable, spreading death and terror. Just over 600 years ago the so-called Black Death alone, in a few years, slaughtered tens of millions throughout Europe and Asia. One third of the population of England was destroyed.

We hope and trust that these horrors cannot happen now. But we sometimes forget that most of the medical discoveries that benefit and protect us have been made only within the last 200 years.

THE ELECTROCARDIOGRAPH. *This is William Einthoven, the Dutch inventor of the electrocardiograph, using his machine. In 1924, Einthoven was awarded a Nobel prize.*

49

The stethoscope

HIPPOCRATES DID NOT USE a stethoscope in the Fourth Century A.D., but he would have found one useful to help him listen more closely to the splashing noise he heard when he shook a patient who had fluid and air in the cavity between the lung and the chest wall. It might also have been more comfortable for the patient to be diagnosed through a stethoscope than by the rough method Hippocrates discovered.

His realisation that it was possible to diagnose certain chest illnesses by ear was important. Unfortunately few of his contemporaries took very much interest and no one who came after him took the idea any further.

Size of organs

It was not until 1817 that a Paris doctor, Rene Theodore Hyacinthe Laennec, developed a simple stethoscope and published a description of it that completely revolutionised the diagnosis of chest complaints.

Laennec had possibly been influenced by a treatise published over 50 years earlier by a Viennese physician, Leopard Auenbrugger, in which Auenbrugger claimed that a doctor could define the size of certain organs and decide whether gas or fluid was present simply by tapping a finger placed on the patient's skin with a finger from the other hand.

Laennec was certainly influenced by contemporary French physicians who studied the sounds produced in the body, particularly the sounds of breathing in the lungs, and heartbeats. One of the most eminent of these physicians was Corvisart de Maret – Napoleon's personal physician – who became an expert on heart disease. It was Corvisart who led Laennec to his own discoveries.

It was also an incident that Laennec noticed in the park that helped: children playing among fallen tree trunks, scraping at one end while their playmates listened at the other. Laennec at once used this idea. He rolled a piece of paper into a cylinder and put one end against a patient's chest; then he listened at the other end and found that he could hear the heartbeats clearly.

Ten-inch cylinder

At first he experimented with wood, cane and several metals for his cylinder, but settled for wood in the end and made a cylinder 10 inches long and $1\frac{1}{2}$ inches in diameter with a quarter inch hole through its length. A number of lengths could be fitted together if necessary.

Later stethoscopes were modified in shape but all were of the one-ear type until 1829, when a London doctor used flexible lead pipes to produce a binaural, or two-ear, stethoscope. This type, however, did not come into general favour in Britain for another 50 years.

Anaesthesia

ANAESTHESIA MEANS TOTAL or partial loss of feeling. It is more than likely that methods of dulling pain were in use in prehistoric times – methods other than the most primitive of hitting someone on

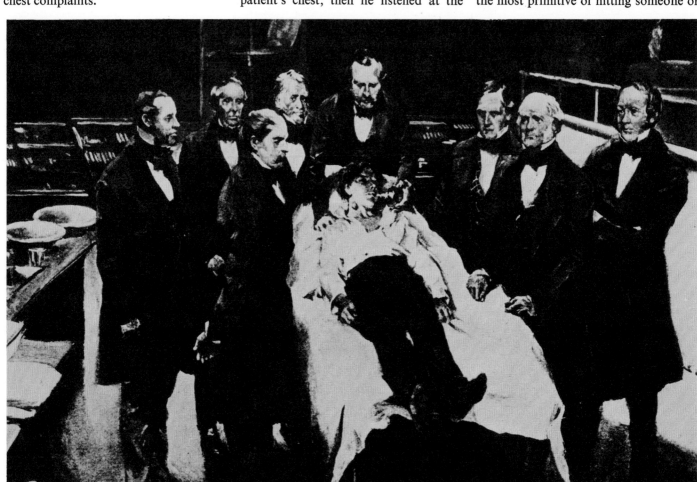

PAINLESS SURGERY. *October 1846. Dr. W. Morton used ether successfully in a major surgical operation for the first time.*

the head in order to make them unconscious!

The Ancient Egyptians probably used the fumes of Indian hemp, which the patient would breathe in; there are references to drugs in Homer and Herodotus; and in the third century A.D. the Chinese surgeon Hua T'o used a wine called *ma-fei-san* for general anaesthesia – his patients were clearly in good hands.

Opium, too, was used widely. In a manuscript from the ninth century there is mention of a soporific sponge which contained a recipe of opium, mandragora, cicuta and hyoscyamus. Similar prescriptions for soporific sponges occur frequently throughout the Middle Ages.

In more modern times Sir Humphrey Davy (who invented the Davy Lamp for miners) suggested that nitrous oxide ('laughing gas') should be used in surgery as it seemed able to destroy pain when it was inhaled.

First ether operation

Nothing was done about this suggestion at the time but 18 years later, in 1818, Faraday discovered that the same kind of effect was produced by inhaling sulphuric ether. This discovery was first put to a practical test in 1844 by Dr. Wells of Hartford, Connecticut, and two years later Dr. W. Morton first used ether successfully in a major surgical operation at Massachusetts General Hospital in Boston.

In the following year chloroform, which had been discovered 16 years earlier, was first used as an anaesthetic for surgery by the great Edinburgh surgeon, Sir James Young Simpson. But it was discovered that chloroform weakened the action of the heart, even though it did produce a deeper state of anaesthesia.

Eventually Trichlorethy and Trilene were found to be more satisfactory than either chloroform or ether.

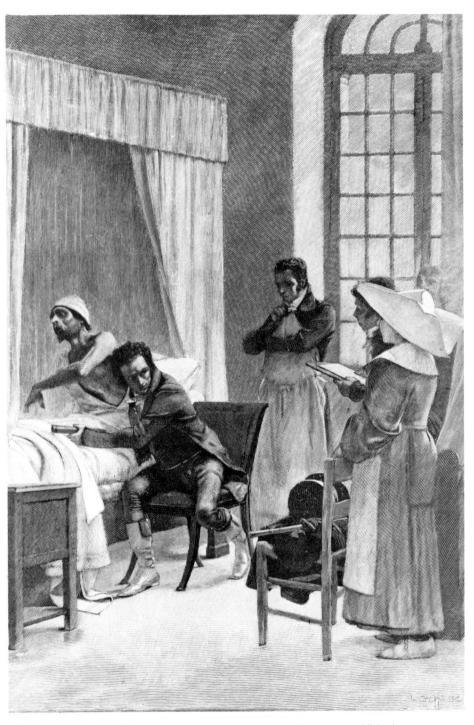

DIAGNOSIS WITH SOUND. *A wooden tube used by Dr. Rene Laennec established the stethoscope as an essential part of every doctor's equipment.*

The X-ray

THE INVISIBLE RAY that could pass through metals and other dense materials was discovered by a German professor of physics, William Konrad Röntgen. He called it X-ray.

He discovered the invisible ray when he was experimenting with vacuum tubes which had been designed by the English physicist Sir William Crookes. Rontgen found that when he passed an electric current through the vacuum tube, a bright fluorescence appeared on an adjacent piece of paper that had been specially treated with a substance called barium platinocyanide.

The fluorescence occurred even when the tube was covered with black cardboard. Rontgen realised that the effect was caused by an invisible ray.

At first doctors used X-rays only to reveal foreign bodies in the tissues and to diagnose bone fractures. But a succession of further discoveries soon encouraged the new science of radiography to delve into the deeper soft tissues of the body. Now radiography has been largely responsible for eradicating tuberculosis from civilised communities and has proved invaluable in cancer research.

The hearing aid

ANIMAL HORNS WERE PROBABLY the first ear-trumpets, but there have been a great many ingenious inventions over the centuries to help the hard of hearing.

Chair-to-chair speaking tubes were used at one time to channel the sound and prevent it from dispersing; the Victorians used mock urns and vases with catchment areas to carry conversation by tube to deaf visitors. But what the deaf really needed was a hearing aid that they could wear, so inventors designed acoustic top hats and acoustic walking-sticks, all with tubes stretching up to the ear. There were several kinds of trumpet used for resonating the sound.

Real scientific advance, however, came when Alexander Graham Bell, whilst struggling to invent a hearing aid, invented the telephone. The principle of the telephone was then adapted to the hearing aid: the carbon microphone turned sound into voltage, which was amplified and turned back into sound. But the batteries were exceedingly cumbersome.

Miniature valves

In 1923, the Marconi Company marketed the valve-operated Otophone – the equipment was packed into a case weighing 16 pounds. About 10 years later, miniaturised valves made possible electronic hearing aids the size of a camera and weighing only four pounds.

The claim to have made the first conveniently wearable aid – weighing two-and-a-half pounds – was made by Edwin Stevens, who founded Amplivox in 1935. But 20 years later, the introduction of the transistor led to the compact size of hearing aid with microcircuits not much larger than a pin-head.

The kidney machine

THE KIDNEY MACHINE WAS A BONUS to mankind from the Second World War, for the Germans were occupying Holland when Willem Kolff developed a device he had been toying with for years and used it in secret to save the lives of his fellow Dutch partisans.

Our kidneys filter waste products from the blood and push them out as urine. When the kidneys stop functioning the waste material accumulates. To deal with this problem the kidney machine 'washes' the blood as it is funnelled through a tube or passed across a membrane.

QUEEN VICTORIA'S EAR-TRUMPET *can contain 374 tiny, transistorized hearing-aids.*

The other side of the membrane is coated with a fluid, and the membrane itself is made of a material which small molecules (like those of salt or water) can penetrate. Such substances pass through the membrane when their concentration is higher on one side than on the other. The excess is thus filtered off from the blood.

Bathing fluid

For instance, blood contains a chemical called urea which is derived from proteins. There is no urea at all in the fluid with which the far side of the membrane is treated – so, the urea passes through from the blood into the fluid as the blood flows through the machine. Obviously by altering the composition of the bathing fluid and by making certain other adjustments any desired quantity of material dealt with by a normal kidney can be induced in or out of a patient's body during treatment.

Kolff's original machine worked well but involved inserting large tubes into an artery or vein. This at first meant a surgical operation every time the machine

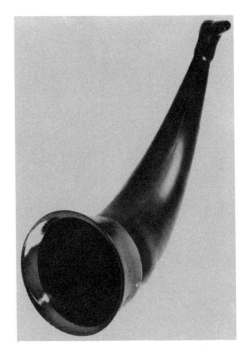

SIMPLE TRUMPET. *This straight-forward ear-trumpet dates from the eighteenth century.*

HIDDEN AIDS. *Many ingenious methods were tried by the Victorians to disguise the hearing aid. Sound receptors could be worn beneath the beard, left, or hidden in a fan, right.*

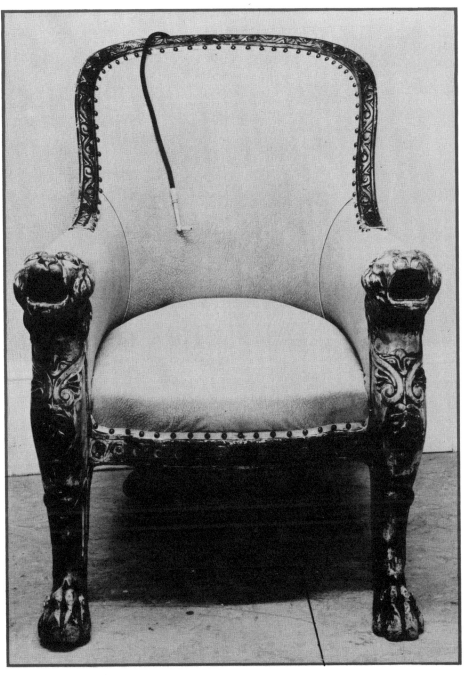

KING GOA'S THRONE. *A Portugese king used this acoustic throne to help his deafness.*

was connected: therefore the machine was most useful in helping acutely ill patients through a crisis.

Permanent tubes

The real value of the kidney machine was not appreciated until after 1960, when Dr. Belding Scribner of Seattle found a method of inserting tubes into a large artery and a fat vein in such a way that they could be left in place for months, even years. It then became possible to treat patients whose own organs had been permanently damaged.

The iron lung

THE IRON LUNG WAS THE first machine to perform a vital function for the human body so that patients who would otherwise have died could be kept alive. It was invented in 1929 by an American, Cecil Drinker, to take over the job of the muscles responsible for respiration.

Chest muscles must contract regularly and sufficiently to enable air to enter. If these muscles fail, quite simply air will not enter the lung – the patient will be unable to breath.

Drinker's solution was equally simple in principle. The patient was put into an airtight box: only the head emerged from the box through a soft airtight collar. The box was then connected to a pump that produced a rhythmical drop in the air-pressure within the box. When the pressure in the box dropped, the pressure of the atmosphere – coming in through the air tubes, throat, nose and mouth – forced air into the chest of the patient: and when the pressure in the box rose against the outside wall of the chest, the air was forced out of the lungs. And so on.

JENNER INOCULATED HIS SON *with swinepox in 1798, a testimony to his own belief in the theory of vaccination.*

A CLOCKWORK DENTISTS' DRILL, *invented in 1864, the first motor-driven device to speed up painful operations.*

The dental drill

THE FIRST DENTAL foot-machine – using a foot treadle for power – was made by George Washington's dentist, John Greenwood, who adapted his mother's spinning wheel to rotate his drill.

Earlier dentists had operated drills with bow-strings, a method that demanded a great deal of courage from the patient and a fair amount of skill and determination on the part of the dentist. Most dentists in Greenwood's day used joiner-type drills which were operated by turning a handle at the side, sometimes with adjustable drill-heads.

Rotary power

The great leap forward came in 1829 when James Naismyth, the Scot who invented the steam-hammer, used rotary power to improve the efficiency of the drill. In the United States Charles Merry patented a hand-operated drill with a flexible cable in 1858, and six years later the first motor-driven drill was invented by an Englishman, George Harrington: it was a hand-held clockwork device with a drill attached.

Not surprising, really, that our fear of the dentist is so deep-rooted!

False teeth

THE EARLIEST KNOWN SET of false teeth was found in a grave at Sidon and dates from the time of the Phoenicians, about 1,000 years before Christ. In this set, the four lower incisor teeth had been replaced by four others, two of which were probably taken from the jaw of another person.

The teeth were not embedded, in fact, they were merely strung together with gold wire, the ends of which were twisted round the two canines.

The Greeks used gold wire to anchor loose teeth, and the Etruscans made gold bridges, surrounding each tooth at its base with a gold ring and adding an extra ring for each of the two natural teeth on either side of the gap. These rings were soldered together to form a continuous bridge. Sometimes human teeth were used, sometimes artificial ones carved from the teeth of an ox.

In recent centuries, Pierre Fauchard, a

Frenchman, who wrote about dental surgery in 1728, first developed the use of the crown, with either natural teeth or ivory. The crown would be fixed to the root of the decayed tooth by a wooden pivot. A session with Fauchard, in his dentist's chair, might last a long time: he took no impressions of the teeth, but preferred to keep on working at the denture until it fitted. Springs were sometimes used to keep the denture in place, but more often than not it relied only on the pressure of the surrounding teeth to stay put.

First gold crown

What was possibly the first gold crown was described by another Frenchman, Pierre Mouton, in 1746. The French chemist Duchateau was the first to make false teeth from porcelain, but he usually made a set at a time; the Italian Fenzi made the first single porcelain teeth.

The electrocardiograph

THE ELECTROCARDIOGRAPH is used to diagnose heart disorders: it records the pulses of electricity that are generated each time the heart beats.

Two German scientists first discovered that the heart of a frog created an electric current, and it was this discovery that decided Willem Einthoven, a Dutchman, to study the electrical activity of the human heart.

Firstly he designed the string galvanometer (a galvanometer is an instrument for measuring electric currents): a thin silver-coated quartz wire is suspended between the poles of a magnet; when an electric current passes through it the wire swings toward a position at right angles to the magnetic line of force. This delicate apparatus is capable of measuring extremely small currents such as they pass through the conducting system of the heart. Furthermore, by placing electrodes on a patient's arm and leg, Einthove discovered that he could detect a pulse of electricity passing through the heart muscles as the blood was pumped around the body.

Beam of light

Einthoven also designed an ingenious way of making permanent records of this current. He arranged the galvanometer string, or wire, so that as it deflected it interrupted a beam of light and produced a shadow on paper. By using a long strip of light sensitive paper and moving it continuously, he was able to produce an electrocardiogram – a continuous record of the heart's muscular activity.

Einthoven designed the galvanometer in 1903. In 1924, three years before his death, he was awarded a Nobel prize.

Vaccination

THE FEAR OF SMALLPOX, the disfigurement it caused and often death, spread throughout Europe in the eighteenth century, and for centuries before had caused pain and terror. There seemed to be no cure, and no one was safe, except, as the old wives' tale went, those who caught cowpox (a disease of the cow's udder).

Edward Jenner, a country doctor from the Gloucestershire village of Berkeley, who had studied in London under the great surgeon John Hunter and from him learnt the value of experiment in medicine, heard of the wives' tale and followed it up. He hesitated for years before daring to inoculate the arm of an eight-year-old boy, James Phipps, with fluis from the blister on a milkmaid's arm, and when he

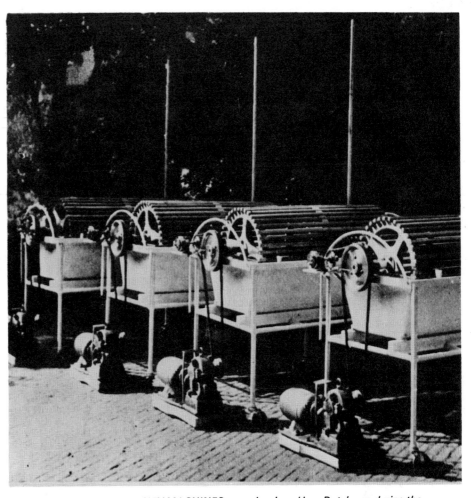

THE FIRST ARTIFICIAL KIDNEY MACHINES *were developed by a Dutchman during the Second World War. But these early machines offered only a short-term hope to patients.*

did the boy developed a similar sore but was not ill.

No reaction

Six weeks later Jenner inoculated the boy again, this time with pus from a smallpox sore, and once again a few months after that. There was no reaction. The mild cowpox infection had protected the boy from smallpox.

Further experiments confirmed this, as well as tests on people who had properly had cowpox. In addition Jenner began a series of arm-to-arm transfers of cowpox fluis and demonstrated that this gave continued protection against smallpox.

He published the results of his experiments at his own expense in 1798 and immediately launched a campaign to persuade other doctors to use the same technique. At first, not surprisingly, he met with hostility and suspicion, but within two years he was voted a Parliamentary grant of £10,000 as a token of his country's gratitude. Before his death, in 1823, he had the satisfaction of knowing

that his technique was in use throughout the world.

Artificial limbs

A VALET'S SON, AMBROISE PARE, was appointed chief surgeon to King Charles IX of France in 1562, at a time when the king was seriously concerned about the vast and growing army of wounded soldiers in his kingdom. Paré is still considered one of France's most eminent surgeons. He was the first man to make artificial limbs acceptable to the medical profession.

He determined to reproduce natural functions with mechanical gadgets, and designed several forms of arms and hands which he fitted successfully to wounded soldiers. His simplest design was an artificial hand which even included a holder for a quill pen; his most complex was a hand in which the fingers were moved separately by a system of tiny cogs and levers.

FRANCE, SEPTEMBER 15, 1916. *A German Army report of that day said: "When the German outposts crept out of their dug-outs in the mist of morning . . . their blood was chilled in their veins. Mysterious monsters were crawling towards them over the craters . . . Someone said, 'The Devil is coming!' Then tongues of flame leapt out of their sides." At the end of the battle, British troops themselves take a look at one of the world's first tanks.*

Part 6

WAR

'War is hell', said General Sherman on his famous march from Atlanta to the sea in the American war between the North and the South. 'War is hell', he said, 'and I intend to wage it as such'. Tales of chivalry and moments of heroism also have their place in war, but only a small place. For the most part Sherman was right: war is fought to be won, by both sides, with any means at their disposal, whether by cunning or by force, so that, with the outmost deliberation, man has progressed in a few thousand years from beating out his opponent's brains with a stone or cudgel to obliterating tens of thousands of his enemies with a single weapon in an instant. Here are some of those methods of force and ingenuity, as man developed them through the centuries, from the cannon to the tank, from the submarine to the Zeppelin, from the machine-gun to the atom bomb – an awesome and deadly array of the instruments of war.

Cannon

THERE IS A STORY that in the early years of the fourteenth century a Franciscan monk called Berthold Schwarz discovered gunpowder and the use of the cannon by mistake whilst working in his laboratory in the town of Freiburg-im-Bresgau.

The story goes that when pounding away at a mixture in his mortar, or basin, with a heavy pestle, the mixture suddenly exploded, blowing the pestle out of the mortar and clear across the room. Fortunately, Berthold himself was not hurt, and when he had recovered from his shock he announced that he had 'invented' both gunpowder and the cannon.

A statue to 'Black Berthold' stands in his home town, but a great many historians now agree that he may never have existed at all. What the historians cannot agree about, however, is precisely who should receive the credit for the development of gunpowder as an explosive charge in cannon and hand-guns. They would like to know, because the idea brought about the greatest revolution in the history of warfare.

It is certain that the Chinese used explosives of some kind as early as 1232, when rockets were used against invading Tartar horsemen, and we can be fairly sure that they had been using explosives for some centuries before that date. Some early chronicles also record that an Arab army that was besieging Mecca in 690 used explosives against the city walls. But explosives at that time were probably used more to scare the enemy by a sudden flash, a loud noise and clouds of evil-smelling smoke rather than to propel a missile.

As for the man who introduced gunpowder into Europe, an English friar,

POT OF FIRE. One of the oldest cannons in the world is "Mons Meg" at Edinburgh Castle. Probably made in about 1450, this massive weapon weighs five tons and has a calibre of twenty inches.

Roger Bacon, holds a better claim to the title than Black Berthold. In 1260 (at least 50 years before Berthold was supposed to have made his discovery), in a treatise on magic and the wonders of nature, Bacon inserted a sentence in a cryptic code which, when deciphered, gave a recipe for making gunpowder by mixing together the correct proportions of saltpetre, charcoal and sulphur.

Cannon captured

Even so, it is still difficult to pinpoint when gunpowder was first used in cannon and what was the first occasion in which cannon were used in battle. Part of the confusion is caused by the fact that the early chroniclers, writing in Latin, used the same words to refer to weapons using gunpowder as they did to catapults and similar weapons.

Various claims for the first appearance of cannon on a battlefield in Europe refer to Amberg, Germany, in 1301; Leydale in Northumberland, 21 years later; Metz in Belgium, two years later still; and Cividale, Italy, in 1331. At Leydale it is recorded that the Scots captured Edward III's artillery, and it is almost certain that they used cannon – possibly the same ones – at the siege of Estrevelin Castle, in 1342.

The first completely reliable mentions of cannon, however, occur in 1326. In that year an order was made by the Council of Florence, in Italy, for 'iron bullets and metal barrels'. And in the same year a certain Walter de Milemete

wrote a book of instructions on kingcraft for the young Edward III of England.

The manuscript is preserved in the library of Christ Church, Oxford, and in it there is a picture of an armoured knight firing a vase-shaped cannon, loaded with an arrow-shaped missile.

The golden colour of the cannon – the type was known as a *pot de fer*, or fire-pot – suggests that it was made of bronze; and if Milemete's drawing is made to scale, we can guess that the cannon was about three feet long.

Cracks filled in

Twenty years later, cannon were in use all over Europe. At first they were made by banding lengths of iron together around a wooden or clay core. Heated metal rings were then forced over the iron bars, binding them together as they cooled and contracted. Any cracks that appeared were filled in with molten metal and then the central core was removed to leave a hollow metal tube.

At one end of the tube, a section was cut away to allow for the insertion of a chamber – a metal, cup-shaped device which held the charge of gunpowder. The stone or metal ball was then dropped into the mouth of the cannon, the train of powder leading to the main charge in the chamber was lit, and the cannon roared.

From the very first, cannon were mounted on timber frames that could be pivotted to bring the muzzle up or down, and so increase or lessen the range. But wheels were not fitted to the frame – to make the cannon a mobile weapon – until the following century.

The earliest cannon still in existence are probably bronze cannon of the late-fourteenth century that were found at Loshult, in Sweden. One of these is very much like the one in Milemete's drawing.

Rockets and missiles

'TAKE ONE POUND of live sulphur, two of willow charcoal, six of saltpetre. Reduce each to a fine powder and mix together. A certain quantity of the final mixture is to be placed in a long narrow cover, and then discharged into the air.'

So run the instructions for preparing a weapon 'to launch fire against the enemy', in a work written by Marcus Graecus in 846 A.D. Although some historians deny its authenticity, others accept it as the first reliable mention of a rocket weapon.

The Chinese and the Arabs used rockets in warfare, and other early mentions of the use of rockets include reports of their use by the Mongols at the siege of Baghdad, in 1258; by the French in the defence of Orleans against the British, in 1429; and by Tipu Sahib against British troops in India, in the 1790s.

However, the first war-rockets to be used in any great numbers and fired by gunners specially trained for the purpose were probably those designed by Colonel William Congreve, at Woolwich Arsenal, London, in 1800-04.

Congreve designed two types of rocket: incendiary rockets, which had a conical metal warhead filled with a slow-burning chemical mixture; and case-shot rockets, which had warheads packed with small shot and an explosive charge. The size of the indendiary rockets varied from 18 to 300 pounds, but the size most often used was the 32-pounder, which had a case about 42 inches long and four inches in diameter, mounted on a 15-foot-long stick. Case-shot rockets were smaller. They generally weighed between three and twelve pounds. Special 'batteries' were made from which to launch the rockets – rather like compartmented metal boxes.

Congreve's rockets were first used in action in 1806, when the British fleet attacked Boulogne, where Napoleon's forces were assembling for the invasion of England. About 2,000 incendiary rockets were fired into the town, setting it ablaze, and confounding Napoleon's plans. The following year, British ships attacked Copenhagen and fired almost 25,000 rockets. Once again, the whole city was set on fire.

The guided missile

ADOLF HITLER'S 'secret weapons' were intended to swing the balance of the Second World War in Germany's favour

at a time when all seemed lost. The V-1 and the V-2 were developed by German scientists, led by Wernher von Braun – later an important figure in America's space programme – at Peenemünde off the Baltic coast. 'V' stands for *Vergeltung*, or 'Vengeance'.

Life-saving rocket

Scientists from many nations had been working for a long time on various aspects of rocketry and missile weapons. In Britain, Colonel Boxer had perfected a two-stage rocket for life-saving purposes in 1855. The Russians, Nikolai Kibalchich and Konstantin Tsiolkovski, later made great advances in the field of rocket-

GUNPOWDER IN THE FRIARY. *Explosive substances were used in China some years before their introduction to Western warfare. The earliest known reference to gunpowder from Europe is the cryptic formula included by Roger Bacon, an English Franciscan Friar, in the treatise on magic and the wonders of nature he wrote in 1260.*

propelled aircraft. And in 1926 the American, Robert Goddard, designed the first liquid-propellant rocket.

However the first true guided missile

rocket was the V-2 (the V-1 had no proper guidance system), and its first successful test was made in October, 1942. In that trial the rocket reached a height of 50 miles in a flight of 120 miles onto a selected practice target.

The first V-2

The first V-2 to reach England was launched from a site near The Hague, in Holland, and fell on Chiswick, in London, on September 8th, 1944. Over a period of about seven months, 1,115 V-2s fell on southern England, killing 2,754 people.

The V-2 was launched from a concrete pad on which it stood upright, balanced on its four-finned tail. It had a maximum range of about 200 miles. The first part of its flight was radio-controlled, so that it was possible to aim at a target with some accuracy.

In flight, the 46-foot-long rocket, weighing about 12 tons with a one-ton explosive warhead, reached a height of 60 miles and a speed of 3,600 miles an hour.

Obviously, no known form of air defence could guard against such a weapon; the only answer was to seek out and destroy the secret workshops and launching pads where it was built and fired. This was done – perhaps only just in time: Von Braun was reported to have been far advanced on a project for a multi-stage rocket that would have been able to strike across the Atlantic at New York, before Allied bombers destroyed his laboratory.

The machine-gun

A MACHINE-GUN IS 'a weapon able to deliver a rapid and continuous fire of bullets as long as the trigger is pressed', but the earliest type of multi-firing gun, sometimes called an 'organ', was simply a row of loaded gun-barrels fixed to a framework, with a trail of gunpowder running to the touch-hole of each.

The gunner put his burning match to the trails and, if the powder was not damp or faulty, all the barrels fired at once in a shattering volley. Sometimes 'organs' were built into small, two-wheeled platforms, to fire through a shield in the front. These were called *ribaulds* or *ribaudequins*.

The first machine-gun worthy of the name was patented by James Puckle, an English lawyer, in 1718. Unlike several *later* guns it had only one barrel, behind which was a cylindrical magazine containing six or more chambers. When the magazine was turned with a crank handle, the shot in each chamber could be fired rapidly, one after the other.

Defending King George

Puckle even provided two kinds of magazine: one held ordinary round shot, for use against 'Christian enemies'; the other held *square* shot, for use against 'heathen Turks'.

Puckle's gun was well in advance of its time but, although it was reported to have fired nine rounds a minute in a trial, no one seemed interested in buying it. Even Puckle's patriotic advertising slogan for the 'Defence', as he called his invention, did not bring the buyers along:

'*Defending KING GEORGE your COUNTRY and LAWES*
'*Is Defending YOUR SELVES and PROTESTANT CAUSE*'.

The American Civil War was a testing-ground for many new weapons – among them the machine-gun. The first use of a machine-gun in action – the 'Williams Gun', of which little is known – is said to have been at the Battle of Fair Oaks, Virginia, in 1862.

It is worth mentioning that in the same year, one of the most famous names in the history of firearms made its first appearance – the Gatling gun.

Ten barrels

Richard Gatling was a planter from North Carolina. His gun consisted of ten barrels revolving round a central axis. The barrels were turned by a crank handle and were fed with cartridges from a 240-round magazine on top of the gun. Each turn of the crank brought a barrel into the firing position and, by moving the other barrels, continued the process of loading or ejecting spent cartridges.

The Gatling could fire from 300 to 1,000 rounds a minute. It was officially adopted by the United States Army in 1866, when Gatling improved the efficiency of the gun by reducing the number of barrels to six.

An air force

MAN HAS NEVER been slow to put new scientific discoveries to military use. So it is not surprising that only about 10 years after the first successful experiments with hot air balloons by the Montgolfier brothers and with hydrogen balloons by Jacques Charles, the world's first 'Air Force' was brought into being. This was the Aerostatic Corps of the Artillery Service, of the armies of Revolutionary France, formed at Paris on March 29th, 1794.

The Aerostatic Corps was commanded by Charles Coutelle, who had perfected a method of generating hydrogen, for filling balloons, without the use of sulphur – all supplies of which were needed urgently for making gunpowder.

Visibility . . . 18 miles

Coutelle's Corps consisted of two officers, a sergeant-major and four NCOs, 25 men and a drummer boy. They wore a blue uniform trimmed with black; on its buttons were a picture of a balloon and the word '*Aérostier*'. But at first they had no balloon. While one was built to Coutelle's own design in Paris, he and his men fought as infantry against the Austrian and Dutch armies that threatened France.

By the early summer of 1794, the world's first military balloon, *L'Entreprenant* ('Enterprising'), was ready for trials. As the two cables fixed to the mid-section of the balloon's silk envelope were payed out, Coutelle, slung beneath, alone in the two-man 'car', rose to a height of 1,770 feet. He found that with a telescope he could observe with ease for about 18 miles. *L'Entreprenant* was set for action.

Under fire

The deflated balloon, together with its brick-built furnace and iron-tubed generator apparatus, and the men of the Aerostatic Corps moved to Maubeuge, in what is now Belgium, where an Austrian army under the Prince of Coburg was deployed against the French. Coutelle went up to make the first observation flight – and immediately came under fire from an Austrian 17-pounder cannon.

'*Vive la Republique!*' yelled Coutelle defiantly. At the same time he signalled hastily to the ground-crew to pay out more cable so that he could rise out of range of the cannon.

The value of Coutelle's observations at Mauberge convinced the French that they had a most important new weapon at their disposal. The Aerostatic Corps was ordered to Fleurus, about 20 miles away, where the French General, Morlot, planned to meet the Austrians in strength.

Because it took nearly two days to inflate the balloon, and longer still to assemble the furnace and the generator, it was decided that the balloon should be moved while it was still inflated. It was far too precious to risk a free flight across the country, so Coutelle and his men, sweating at the cables, had to tow the unwieldy monster along like a giant kite.

General aboard

On June 26th, 1794, the Austrian soldiers prepared for battle. They had been told by their officers that the French Revolutionary Army was 'in league with the Devil himself' – and many of them must have believed it when they saw *L'Entreprenant* rising over the French lines.

Coutelle was in control of the balloon, signalling to the ground-crew with flags. With him was General Morlot, who scanned the Austrian lines through a telescope. By watching the movements of the enemy troops below, Morlot was able to judge the intentions of the Austrian commanders, and by passing messages to his own officers by way of a bag and cord fixed to the balloon's cables he was able to counter every Austrian move and win a decisive victory.

Daring exploits

However, the Aerostatic Corps did not last long, in spite of its success. Napoleon, who became commander-in-chief of the French army shortly after Fleurus, did not think that war should be waged with balloons. For one thing, they detracted from the military glory of his own individual genius.

It is said that Napoleon was not the only French soldier to look with disfavour on the Aerostatic Corps: the daring exploits of the *Aérostiers* made them so popular with the girls wherever the army went that no other soldiers stood a chance.

Whatever the reasons, the Aerostatic Corps was disbanded in 1799, only five years after its foundation.

The Zeppelin

COUNT FERDINAND VON ZEPPELIN, creator of the world's largest dirigibles, or airships, made the first flight in LZ1 – the first of the airships to bear his name – in 1900. By the time the Great War broke out fourteen years later, the German High Command had been persuaded that in the Zeppelin they had a potential war-winner.

The Zeppelin's supporters pictured British cities shattered and smoking after bombing raids, with terrified citizens storming the Houses of Parliament to demand peace at any price. On the outbreak of war, the German Naval Air Service was ready with a fleet of Zeppelins to bomb Britain into surrender.

However, things did not go quite as planned. It was one thing to manoeuvre a Zeppelin in peacetime, in daylight, in fine weather; it was quite another to take the same craft several hundred miles on a dark, stormy night, with all its lights shielded against enemy planes or fire from the ground, in order to drop a heavy bomb load on a selected target.

Target ... London

The first bombing raid on England took place on 21 December 1914, when a German aircraft dropped two bombs which fell in the sea near Dover pier. There was a second raid on 24 December; on this occasion a bomb exploded near Dover castle. Three airships took part in the raid on 19 January 1915, only one of which had orders to attack London.

Kapintanleutnant Hans Fritz, commanding L3, dropped his bombs on Yarmouth, on the coast of Norfolk. In the same area, L4 claimed the first civilian air raid casualties in a raid on Britain; its bombs fell on the villages of Snettisham and Sheringham, killing two people and injuring 13.

Next day, the raid was headlines all over the world: 'ZEPPELINS BOMB SAND-RINGHAM AS KING GEORGE AND QUEEN FLEE: PANIC GRIPS LONDON', screamed one American newspaper. But far from terrifying the British, the Zeppelin's killing of civilians made them furiously angry. They determined to hit back.

It was this determination to hit back against Zeppelin attacks that brought about the first Zeppelin war casualty. The incident was a dramatic one.

Flight Sub-Lieutenant Reginald Warneford of the Royal Naval Air Service took off on a bombing mission at one o'clock in the morning in June, 1915. He was an experienced pilot but had never before flown at night. He had left Furnes airfield, on the French-Belgian frontier, when he met three Zeppelin airships returning from a raid on England.

Over Ostend, he saw the pale outline against the night sky of a cigar-shaped monster that seemed, in his own words, 'to be about half-a-mile long'. It was Zeppelin LZ37, commanded by Oberleutnant van der Haegen.

Warneford's tiny aircraft carried no guns, but in a rack beneath it, ready to be released by a simple toggle-and-cable device, were six 20-pound Hale Naval grenades – small incendiary bombs. The huge airship, with its four 210 horsepower Maybach engines and its Parabellum machine-guns fixed in the gondola beneath the envelope began a strange duel with the small plane in the night sky over Belgium. Warneford's only hope was to get above the Zeppelin and drop his bombs on it.

The odds should have been on the Zeppelin, with its faster rate of climb, but somehow, above Bruges, at a height of about 7,000 feet, Warneford found himself on top, in position for a bombing run and out of the line of the machine-guns' fire. One hundred feet below him the 521-foot-long, green-painted back of the airship offered a massive target. Warneford let all six bombs go together in a single 'stick'.

The next thing that he knew was that a huge explosion was hurling his Morane around the sky. His bombs had ignited the 953,000 cubic feet of hydrogen in the Zeppelin's envelope – LZ37 was already a flaming mass of wreckage, plunging toward earth with its 28-man crew. Only one survived: the helmsman leapt from the falling gondola at about 200 feet, crashed through the roof of a nunnery, and landed in an unoccupied bed with only a few bruises.

Warneford fought his plane back on an even keel, only to hear the engine cut out, leaving him to glide down in pitch darkness many miles behind the German lines. He landed safely and, afraid to show a light, set to work on his engine with only his sense of touch to guide him. Amazingly, he discovered a simple break in the petrol feed pipe; he bandaged it with his handkerchief, took off again, and landed safely behind his own lines.

Thirty-six hours after his historic victory, Warneford learnt that he had been awarded the Victoria Cross. He was ordered first to Paris, where the French Government were to present him with the Legion of Honour.

'What rejoicing there will be when you get back to England,' remarked a Frenchman. Without a trace of emotion, it is said, Warneford replied, 'I do not think I shall make it.'

He was right. A few days later, he was testing a new Henri Farman biplane at Buc airfield, near Paris. Just after take-off the aircraft turned on its back and crashed. Warneford and his passenger, an American journalist, were killed instantly.

CAPTAIN COUTELLE'S "AEROSTATIC CORPS". *An observation balloon ensured French victory at Fleurus in 1794.*

ROLLS ROYCE ARMOURED CAR. *Prototype armoured vehicles used in the Great War.*

H.M.S. ARGUS. *The first true aircraft carrier was launched in 1917, converted from an unfinished liner.*

FIGHTER ACE. *By 1916 the German pilot Oswald Boelcke had the best claim to be the first "ace".*

VE. ZEPPELIN DESTROYED. *June 1915.*
t Sub-Lieutenant Warneford of the Royal
bombed a German airship.

ABOVE. TOTAL WAR. *August 1945. The Japanese city of Hiroshima was torn apart by the*
holocaust of an atomic explosion. A single bomb devastated the area, killing or severely
injuring 90,000 people.

IAN TORPEDOES ATTACK GIBRALTAR. *September 1941. Two Italian frogmen steered their tiny submarine into the heavily guarded*
our and sank a tanker before escaping to neutral Spain.

Aerial combat

WITHIN A FEW YEARS of the Wright brothers' first flight in 1903, the aeroplane was being put to the test as a weapon of war. Aircraft were first used for military purposes in the Italo-Turkish war of 1911-12.

In November, 1911, Lieutenant Giulio Gavotti became the first 'bomber' pilot when, flying over a Turkish position at Ain Zara, in Libya, he tossed over the side of his cockpit a four-pound explosive charge.

Combat between two aircraft did not take place until two years later. During the Mexican Civil War, the Americans Phil Radar and Dean Lamb, making reconnaissance flights for opposing sides, fought a brief duel with pistols – neither inflicted any damage on the other.

Speed . . . 60 m.p.h.

The nations involved in the Great War started off with very few aircraft: France had 136, Russia had 224, Germany had 198, and Britain had only 63. The first fighting planes were flimsy contraptions of 'sticks-and-string', with a maximum speed of 60 miles an hour. At first they carried no guns: the pilots were armed only with pistols, rifles and – on at least one occasion – house-bricks on lengths of rope, with which they hoped to foul their opponent's propellers.

One British 'expert' suggested that county cricketers should be recruited to fly as passengers, armed with hand-grenades which they would lob with unerring accuracy into enemy cockpits. But from these unpromising beginnings there evolved in only four years of war properly organised 'air forces', specialised fighter and bomber aircraft, and a new kind of fighting man – the 'ace', or champion fighting pilot.

The fighter 'ace'

SEVERAL MEN HAVE a claim to be considered the first air 'ace'. Many authorities feel it was the French pilot Roland Garros. But, among the contenders are the British Captain, A. A. B. Thompson, who is credited with originating the technique of 'ground-strafing', or machine-gunning enemy troops from the air, and the German Max Immelman, originator of the manoeuvre known as the 'Immelman turn', a sharp dive and side-slip to bring the pilot into an attacking

position on the unprotected underside of the enemy aircraft.

But perhaps the best claim is that of Oswald Boelcke, the guiding spirit behind the formation of the first 'flying circus'.

Boelcke was 23 at the outbreak of the Great War, a small, rather shy young man who suffered severely from asthma. After gaining a place in the air service, he made about 40 uneventful flights in an unarmed biplane 'spotting' for German artillery batteries. Early the following year he was transferred to a formation of Albatros 'fighter' aircraft, which carried a machine-gun mounted on the lip of the rear cockpit.

Burst of fire

On July 6th, 1915, Boelcke made his first kill. He stalked a French Morane Type L Parasol two-seater, also armed with a machine-gun, for about 30 minutes before gaining an attacking position just above it. The French pilot avoided his burst of fire by putting his machine into an 80-mile-an-hour dive. For a further 20 minutes the two aircraft kept up a running battle 2,000 feet above the German lines, until the Morane, riddled with machine-gun fire, went into a spin and crashed. Boelcke landed nearby and found that both the French pilot and gunner were dead.

Although he scored 40 victories, Boelcke was always anxious that his victims should survive: he would make a point of visiting them, in hospital or prison, with gifts of wine and cigarettes. Even after winning the 'Pour le Mérite' ('The Blue Max'), his country's highest honour, he is said to have gained more satisfaction from the award of a life-saving medal he received after plunging into an icy river to rescue a drowning boy.

'Fly close'

Boelcke was a lone wolf, but even so it was he who was asked to form *Jagdstaffel*, a squadron of crack pilots, picked by himself and equipped with the most efficient aircraft available, which the British later described as a 'flying circus' (because of the German habit of painting their aircraft in bright colours for identification purposes).

To the young pilots under his command, Boelcke taught his simple method of combat: 'I fly close to my man, aim well, fire, and then he falls down!' But he still kept the habit of taking off and hunting on his own. A pilot who served with him explained what happened when

he returned from a mission. The other pilots would ask what success had he had. Climbing from his aircraft, Boelcke would ask in turn: 'Is my chin black?' If it was, grimy with powder from firing his Spandaus, then it was certain that he had made a kill.

The twin Spandau machine-guns gave him his greatest successes. They were fitted to the revolutionary Fokker DIII biplane, and by means of the interruptor-gear designed by a Dutchman, Antony Fokker, they were able to fire through the propeller of the aircraft. This meant that the pilot himself could operate the gun, and opened the way for single-seater fighters – ideal for Boelcke's individual talent.

Boelcke's death

His death, however, did not come, as he might have wished, in single combat with an enemy aircraft, but as the result of an accident. It is described by one of his most promising pupils in a letter written a day or so later:

'*Boelcke, some men of our squadron, and myself were engaged in a battle with British planes. Suddenly I saw how Boelcke, while attacking the enemy, was rammed by one of our gentlemen, to whom, poor fellow, nothing else happened. I followed Boelcke immediately. But then one of his wings broke away and he crashed down. His head was smashed by the impact: death was instantaneous.*

'*In the last six weeks we have had twelve of our pilots shot down, six dead and one wounded, while two have suffered a complete nervous collapse. Yesterday I brought down my seventh enemy aircraft. The ill luck of the others has not yet affected my nerves.*

'*With my love, dearest Mamma, Manfred von Richtofen*'.

An airborne invasion

THE IDEA OF THE PARATROOPER as a kind of 'superman', ready to strike from the sky at a moment's notice, had great appeal for the German dictator Adolf Hitler. 'This is how the wars of the future will be fought,' he wrote, 'the sky black with bombers, and from them, leaping into the smoke, the parachuting storm-troopers, each one grasping a sub-machine gun.'

The parachute had first been used in offensive warfare in 1939, when Russia invaded Finland. About 200 paratroops of the Red Army landed successfully from six aircraft near the village of Salmijärvi. The German army was quick to follow the tactic: paratroopers were used in limited numbers, as shock-troops or saboteurs, in their invasions of Norway, Holland and France in 1940.

But it was when the Germans faced the task of capturing the Mediterranean island of Crete that the time had come for the first full-scale airborne invasion in military history.

Battle-weary men

Crete offered an ideal airbase for the control of the Mediterranean. It was garrisoned by 25,000 Allied troops, mostly Australians and New Zealanders, under the command of General Freyberg, V.C. Most of Freyberg's men were battle-weary after being evacuated from Greece under the noses of the advancing German army. Their supplies of food and ammunition were low; they had few anti-aircraft guns and only a handful of armoured vehicles.

The Germans had total air superiority; the Royal Navy controlled the sea. Airborne invasion was the obvious answer to the Germans' problem: Operation *Merkur* (called '*Scorcher*' by the Allies) was put into action.

Against Freyberg's ill-equipped defenders, General Kurt Student planned to throw 'the toughest fighters in the German army' – a full parachute division, a division of mountain troops, an armoured regiment, a motorcycle battalion, and light artillery units. To carry his invasion force, he assembled more than 500 Junkers Ju52 bomber-transports, each capable of carrying 4,000 pounds of supplies or fifteen fully-armed men, and a number of heavy gliders to be towed behind the transports, carrying more men, light vehicles and artillery.

Pistol and knife

To cover his transport planes, he had a large fleet of fighters and bombers, including about 150 Junkers Ju87 'Stuka' divebombers, commanded by Wolfram von Richthofen, younger cousin of the famous 'ace' pilot of the Great War.

Each German paratrooper, equipped for battle, wore a leather jerkin over his uniform, a crash-helmet, and a camouflage cape. His feet and legs were protected against the shock of landing by half-length rubber-soled boots, and his body by pads built into his parachute harness at chest, knees and shoulders. He

SUPERMEN FROM THE SKY. *In 1940 the German Army pioneered the use of paratroops as a spearhead of their attack on Holland.*

carried with him on his drop only a pistol and a long knife: heavier weapons were dropped in separate containers.

After landing, each combat section consisted of one *Solothurn* light-machine-gun crew, eight men armed with *Schmeisser* sub-machine-guns, and two marksmen armed with *Mauser* rifles with telescopic sights. Included in every drop were sections armed with anti-tank weapons, flame-throwers and light and heavy mortars. Each man carried concentrated rations to last two days and 'pep-pills' to take if he felt tired.

Easy targets

The first wave of aircraft swept over Crete on the morning of May 20th, 1941.

To the island's defenders, the paratroops looked like 'balloons coming down at the end of a party' – the parachutes were brightly coloured: violet or pink for officers, black for other ranks, yellow for medical supplies, and white for arms and ammunition.

It seemed at first as if the invasion would fail: the paratroops made easy targets whilst in the air and immediately on landing; the clumsy gliders were riddled by bullets in the air, or crash-landed, killing many of their occupants; attempts to reinforce the invaders by sea were broken up by the Royal Navy. But over ten days, as more airborne troops were flown in, the Germans gradually gained the upper hand.

Their fighters controlled the sky; their divebombers blasted the defenders' positions at will; the Allied troops ran short of ammunition. One New Zealand Maori soldier broke up several German attacks on his position by throwing hand-grenades. When his supply ran out, and the Germans advanced again, he charged forward throwing grenade-sized stones instead. The Germans retreated without waiting for the 'grenades' to explode.

Germans salute

The Cretan villagers joined the heroic hand-to-hand fighting with ancient muskets, axes and pitchforks, while isolated Allied forces fought a rear-guard action and the Royal Navy staged a minor 'Dunkirk', under heavy air attack, to take off as many soldiers as they could. Naval losses were many, including the cruiser *Gloucester*.

'*She went down with all her guns blazing upwards, and we cheered her from the cliffs,*' wrote an Australian soldier later, '*and when she'd sunk, the German pilots flew over and dipped their wings in salute.*'

Between May 27th and 30th, the remaining Allied forces surrendered, although some took to the mountains and fought as guerrillas for many months. Total Allied losses were estimated at 15,000, killed, wounded or captured; the Germans lost about 6,000, dead and wounded – but many of these were their finest troops, and the German High Command became wary of again using paratroops as a major attacking force. In fact, Crete marks the first and almost sole occasion (with the possible exception of the Allied landing at Arnhem in 1944) on which paratroops have been used in such a role.

The Ironclad

THE AMERICAN CIVIL WAR was not the first conflict in which steam-propelled, armour-protected warships played a part.

The first steamship to go into battle was probably the East India Company's sloop *Diana*, which was used as a gunboat off Burma, in 1824, by Captain Marryat – better remembered as the author of the adventure novels 'Midshipman Easy' and 'Peter Simple'. Also, in the Crimean War, French steam-frigates with protective armour-plating were used to shell the Russian fortresses.

What the American Civil War *did* see

was the first battle between armoured ships, or ironclads, and the first use in naval warfare of the revolving gun-turret.

Six guns

Merrimack began her life as a 3,200-ton steam-frigate of the United States Navy, built in 1855. Six years later, when the armies of the Confederacy, the southern states, seized the Navy Yard at Norfolk, Virginia, the retreating northerners, the Federals, destroyed as much of the Yard as they could before withdrawing. Among the ships they damaged was the *Merrimack*, which was burned down almost to the waterline.

The Confederates trimmed off the sides of the captured wreck and decked her over within three feet of her waterline. On this floating platform they built a long deck-house with sloping sides, tapering toward the roof, pierced in order to take two eight-inch Dahlgren rifled guns and four smooth-bore cannon.

Both the deck and the deck-house were protected by iron railway-lines bolted on in two layers; the upper part of the funnel, sticking up through the deck-house roof, and a small triangular conning tower, also on the roof, were protected in the same way. At the bow, a great iron ram jutted forward – the *Merrimack* combined one of the oldest methods of sea-warfare with one of the newest.

Into attack

Her appearance was summed up neatly by a Federal officer who saw her in action: '*A very big barn, belching out smoke and flame*'.

On March 8th, 1862, *Merrimack* – she was rechristened *Virginia* but is still best-remembered by her original name – steamed down the Elizabeth River to attack the Federal fleet in Hampton Roads. Her commander, Commodore Buchanan, and her crew of 350 men were in high spirits: in spite of her poor engines, her faulty steering gear, and her draft of 23 feet, which made her hard to handle in shallow water, *Merrimack* was the most powerful ship afloat.

Disregarding the fire of a 30-pounder cannon from the Federal armed tug *Zoave*, *Merrimack* steamed toward *Cumberland*, *a* 24-gun sloop, and *Congress*, a powerful 50-gun frigate.

The shots from these two ships simply bounced off *Merrimack*, even at point-blank range. Badly damaged by *Merrimack*'s own broadside, *Congress* soon ran

aground, where she was captured an burned by Confederate troops, whil *Cumberland* was sunk by *Merrimack* ram.

A third Federal ship, the 40-gun frigat *Minnesota*, was also badly damaged an ran aground.

Commodore Buchanan was wounde in the action. He handed over his com mand to Lieutenant Catesby Jones an ordered him to finish off *Minnesota* in th morning. But *Merrimack*'s command the sea did not last for another day.

The second ironclad

While she lay anchored, at night, second ironclad entered Hampton Road – the 124-foot-long *Monitor*, which ha been specially built for the Federals b the Swedish engineer John Ericsson. Sh carried only two guns, but they wei

1-inch Dahlgrens, the most powerful naval guns then in existence. And they were mounted on a revolutionary device – an iron turret, nine feet high and 20 feet round, weighing 140 tons, revolving round a brass rail set into the vessel's keel. Here, for the first time, was a warship that had no need to manoeuvre to bring her 'broadside' of guns into action.

In appearance, *Monitor* was nothing more than an iron-plated raft, almost level with the water, supporting the gun-turret, a tiny five-foot-high pilot-house, and two short funnels.

Guns roared

In the morning, *Merrimack* weighed anchor and set out to finish off the crippled *Minnesota*, from behind which *Monitor* suddenly steamed to give battle. As the guns of the ironclads roared out,

the remaining warships in Hampton Roads sheered away. The long reign of the 'wooden walls' that had made up every fleet from Salamis to Trafalgar was over: the days of iron and steel had begun.

For more than two hours the ironclads blazed away at each other. *Monitor* could fire both guns at seven-minute intervals; *Merrimack* delivered a broadside approximately every 15 minutes. But neither ship could damage the other.

'I can do as much damage by snapping my fingers,' reported Merrimack's gunnery officer. But when Lieutenant Jones decided to break off the action and concentrate on the helpless *Minnesota* he ran aground. Further attempts to close with *Monitor* and ram her also failed.

The historic battle ended in a draw, when *Monitor's* captain was hit by a lucky

MERRIMACK AND MONITOR. *March 1862. The first battle between ironclad ships took place in American Civil War. For two hours the Confederate Merrimack and the Union Monitor blazed ineffectually before they both withdrew.*

shot on the pilot house just as *Merrimack* was once again in danger of running aground as the tide ebbed. Both withdrew, with the honours even.

Neither *Merrimack* nor *Monitor* survived the year of their glory. On May 11th, 1862, the Confederates scuttled *Merrimack* at Norfolk – where they had found her – this time to prevent her capture by the advancing Federals. As for *Monitor*, she sank in a gale off Cape Hatteras, on December 31st, with the loss of 16 officers and men.

The submarine

'Its shape was most like a round clam, but longer and set up on its square side,' explained Sergeant Ezra Lee, a soldier in George Washington's army during the American War of Independence, when he tried to describe *American Turtle*, the first operational submarine craft.

The Americans were sorely in need of a 'secret' weapon in 1776, for the British, helped by their command of the sea, were pressing hard on New York, and Washington's army was being daily thinned by desertions. It was not surprising, therefore, that when the Connecticut inventor, David Bushnell, offered his strange new craft to Washington, it was accepted at once. And since Bushnell himself was not strong enough to operate the submersible, a volunteer was found in the person of Sergeant Lee.

Lee later wrote a full description of *American Turtle*. The vessel was just large enough for a single man to stand or sit in, with six tiny windows of thick glass, 'each the size of a half-dollar piece', giving him enough light 'to read in three fathoms of water'. Not that he would have time for reading: with one hand he turned a crank which drove the craft along by means of two short, broad paddle-oars; with the other he operated a tiller leading through a waterproof joint to an external rudder.

A pocket compass

'With hard labour,' wrote Lee, 'the machine might be impelled at the rate of three knots an hour for a short time.'

To submerge the *Turtle*, Lee had to press down a spring-operated water vent with his foot. Water flowed in, adding its weight to the 700 pounds of lead ballast, and the craft gently sank beneath the surface. If it sank too far, the operator had to work hard at the hand-pumps beside his seat in order to force water out again. His 'instrument panel' for underwater navigation consisted of a pocket compass and a cork floating in a sealed glass tube – a primitive depth-indicator.

Fixed to the *Turtle's* upper side was a sharp screw, worked by a crank inside the craft. Having manoeuvred the submersible beneath an enemy ship, the operator had to drive the screw firmly into its bottom and then release a waterproof explosive charge which was attached to the screw by a line. A clockwork timing mechanism on the charge allowed the submarine time to get clear before the bottom was blown out of the unsuspecting enemy vessel.

It was a fine night early in September, 1776, when the *Turtle* first went into action. Rowing boats towed her, with Sergeant Lee inside, to a point in the East River just above Staten Island, where the British fleet lay at anchor. Cast off by the boats, Lee found himself obliged to paddle hard to prevent the tide from carrying him away from his target, but at about one o'clock in the morning he came alongside the British man-of-war *Asia*. What happened then is best told in his own words:

'When I rowed under the stern of the ship, I could see men on the deck and hear them talk. I then shut down all the doors, sunk down, and came under the bottom of the ship. Up with the screw against the bottom, but found that it would not enter. I pulled along to try another place, but deviated a little to one side and immediately rose with great velocity and came above the surface two or three feet between the ship and the light, and then sunk again like a porpoise.'

Retreat . . .

'I hove about to try again . . . but then I thought the best generalship was to retreat as fast as I could, as I had four miles to go before passing Governor's Island . . . . When I was abreast of the Island, three or four hundred men got upon the parapet to observe me. At length (they) shoved off a twelve-oared barge and pulled for me.

'When they got within fifty or sixty yards of me, I let loose the magazine in hopes that if they should take me, they would likewise pick up the magazine and then we should all be blown up together. But as kind Providence would have it, they took fright and returned to the island. . . . Our people seeing me, came off with a whaleboat and towed me in. The magazine, after getting a little past the island, went off with a tremendous explosion, throwing up large bodies of water to an immense height.'

1800 . . . the Nautilus

So Lee's brave attempt failed, and no further experiments were made with *American Turtle*. It was left to another inventor, also in America, to carry on Bushnell's work: Robert Fulton designed his *Nautilus* in about 1800. And it was not until nearly a century after Fulton's experiments that the first submarines of the 'Holland' type, again American-designed, entered service with the world's navies.

Human torpedoes

ON THE MORNING OF September 20th, 1941, part of the British Mediterranean fleet – including the aircraft-carrier *Ark Royal* and the battleship *Rodney* – lay at anchor in the Grand Harbour of Gibraltar. Every precaution had been taken to guard against Italian or German submarine attack. Anti-submarine netting of heavy metal mesh was strung across the harbour entrance, and destroyers patrolled outside.

Early that morning, the netting was lowered to allow one of the patrolling destroyers to re-enter Grand Harbour. Silent and unseen beneath the surface there crept in at the same time a slim cylinder, just over 20 feet long, with two men perched astride: Lieutenant Visintini and Petty-Officer Magro of the Italian Navy. For the first time, the 'human torpedo' was about to strike a successful blow in naval warfare.

Avoiding *Ark Royal*, which they knew would be heavily guarded, the two 'charioteers' (as they were called) chose the big tanker *Denby Dale* as their target. With their twin propellers pushing them along at a snail's pace – the torpedo's maximum speed was only three knots – they brought the 'pig' (the name the Italian seamen gave to their awkward craft) beneath *Denby Dale's* hull.

Fuse set

Their tight rubber suits and clumsy breathing gear did not allow the 'charioteers' to work very fast. They had to clamp lines to *Denby Dale's* keel and from them suspend the torpedo's detachable 500-pound explosive warhead just under the ship's bottom. Once the delayed-action fuse was set, Visintini and Magro cleared out as fast as they could to neutral Spain, where an Italian sabotage group was waiting to pick them up and whisk them back to a heroes' reception in Italy.

At 8.43 am the explosive charge under *Denby Dale* went off. Minutes later, the tanker was a wreck on the harbour floor. Hardly had the noise subsided when, outside Grand Harbour, where two more 'pigs' had been at work, similar explosions seriously damaged the 10,900-ton freighter *Durham* and the small tanker *Fiona Shell*. The human torpedoes could claim resounding success.

Earlier attempts had failed. The idea for a human torpedo was first put forward six years earlier, when two Italian officers

The first underwater battle

Sub-Lieutenants Toschi and Tesei, drew up a blueprint. After experiments with a prototype in the same year, the Italian Navy set up 'H. Group', for research and training in underwater warfare. By the time Italy entered the Second World War as Germany's ally in 1940, 'H. Group' had become part of the Tenth Light Flotilla, a force specialising in the use of such shock weapons as EMBs (one-man explosive motorboats), free-swimming 'frogmen' and human torpedoes.

In August, 1940, a submarine ferrying three 'pigs' toward Alexandria was sunk by a British aircraft, and in the next month an attack on Gibraltar was abandoned when it was realised that no large ships were in the harbour. More disaster for the torpedoes occurred off Alexandria a little later: the submarine *Gondar*, carrying 'pigs' and their crews, was depth-charged, and Lieutenant Toschi himself was captured.

Two broke down

There were mechanical failures as well. In November, 1940, when three 'pigs' struck at Gibraltar, two broke down and were scuttled by their crews. The third reached its target, the battleship *Barham*, but then the crew's breathing apparatus failed. The two men scuttled their craft, surfaced and were captured.

By this time British Intelligence had some idea as to what the Italians were up to. In the summer of 1941, after further Italian attacks against Gibraltar and Malta had also failed, the British set up a research and training group called 'Under Water Working Party', led by Lieutenants William Bailey and L. P. K. 'Buster' Crabb. This group was responsible for developing many of the techniques later used successfully by British 'charioteers' and 'frogmen'.

Triumph at last

The greatest triumph of the human torpedoes came almost exactly three months after the incident in Grand Harbour. On December 18th, 1941, three Italian 'pigs' penetrated Alexandria harbour and succeeded in badly damaging the battleships *Queen Elizabeth* and *Valiant*.

There was a strange sequel to the attack. Count Luigi de la Penne, who had led the attack on *Valiant*, later became one of the Italian underwater experts who worked with the British after the collapse of Italy in 1943. In March, two years later, he was awarded the *Medaglia d'Oro*,

Italy's equivalent to the Victoria Cross. The medal was pinned to de la Penne's breast by Vice-Admiral Sir Charles Morgan, who had been Captain of *Valiant* when de la Penne had so nearly sent her to the bottom in Alexandria harbour.

The Victoria Cross

ON JUNE 26TH, 1857, 62 very unusual men stood to attention in Hyde Park, London. All soldiers or sailors, they were parading at the command of Queen Victoria, to receive from her own hands the medal she had founded the year before, the highest award any British serviceman can receive – the Victoria Cross. To one of the men a special honour belonged: Mate Charles Lucas was the first to win the new medal.

Almost exactly a year before, on June 21st, 1854, Lucas was on the deck of *HMS Hecla*, in the Baltic Sea. Britain was at war with Russia, and British warships had been sent to bombard the powerful fortress of Bomarsund, in what is now Finland. For the wooden ships of the time, an engagement with a shore battery was extremely dangerous: the gunners on the land used furnaces in which shot was made white-hot before firing.

Fizzing fuse

It was one of these shells that fell into the middle of the group of men to whom Lucas was giving instructions. Furthermore, the shell – a hollow steel ball – contained hundreds of tiny steel pellets: when its fizzing fuse burned down to an explosive charge at the shell's core, the ball would burst and the small shot inside would sweep the deck. As it landed, the fuse had already almost burned down – it was a matter of seconds.

Pushing his men aside, Lucas snatched up the shell, still glowing red-hot, ran to the ship's side and hurled it overboard. It exploded even before it hit the water.

For this act of *exceptional personal bravery in the presence of the enemy*, not only did Lucas receive the Victoria Cross but he was also promoted immediately to Lieutenant. He lived to become an Admiral.

Russian metal

Victoria Crosses are made of bronze from Russian guns captured at the battle of Sebastopol in 1855.

Since 1918 the medal's ribbon, instead of being blue for the Navy and red

for the Army, has been universally claret coloured. Otherwise, the Victoria Cross has remained unchanged since Lucas received his. The simple Maltese cross, one and a half inches across, bears in its centre the Royal Crown surmounted by a lion. Below the crown, on a semi-circular scroll, are the words: FOR VALOUR.

Seaplanes

GLENN HAMMOND CURTISS, the man who designed the plane in which Commander Ely made his first successful take-off and landing on a warship, was himself the first man to make a successful flight in a seaplane. On February 17th, 1911, he landed his plane on the water beside a naval vessel, on to which he was then hoisted by crane.

The first successful seaplane bombing raid was made by aircraft similar to the Short S.81, Type 184 seaplane in the illustration, powered by a 160 horsepower Gnome engine and mounting a one-and-a-half-pounder quick-firing gun in the front cockpit.

In 1914, Short seaplanes from the seaplane-carriers *Empress*, *Engadine* and *Riviera* – all converted cross-Channel steamers, fitted with cranes and carrying four aircraft each – bombed German Zeppelin sheds at Cuxhaven.

The first seaplane-carrier to go into action was a 7,500-ton converted collier. In 1914, the Royal Navy fitted her with cranes to lift aircraft from her hold to the surface of the water, and rechristened her with a famous name – *Ark Royal*.

Early seaplanes did not carry torpedoes although a 'torpedo plane' had been patented by an American, Rear-Admiral Bradley Fiske, in 1912. The first recorded sinking of ships by aerial torpedoes occurred on 12 August, 1915, when Flight Commander C. H. Edmonds, flying a Short Type 184 seaplane, torpedoed and sank a Turkish supply ship in the Sea of Marmara. This ship has been disabled by the British submarine, E14, a few days earlier. On 17 August, Edmonds and Flight Lieutenant G. B. Dacre sank two other Turkish steamers.

Aircraft carriers

THE FIRST AIRCRAFT to take off from the deck of a warship was a Curtiss biplane flown by Commander Eugene B. Ely. On November 14th, 1910, he took off from a

83-foot-long wooden deck built over the bows of the U.S. Navy's cruiser *Birmingham*, at Hampton Roads, Virginia – where the historic battle had taken place between *Merrimack* and *Monitor* nearly fifty years earlier.

In January, the following year, Ely became the first pilot to *land* an aircraft on a warship. He put his biplane safely down on the specially-prepared after-deck of the *U.S.S. Pennsylvania*.

The first catapult-assisted take-off, from a compressed-air catapult fitted to a barge in the Potomac River, was made by Lieutenant Ellyson of the U.S. Navy, in 1912.

The first successful take-off and return landing on a ship actually under way was made from the converted cruiser *H.M.S. Furious*, in 1917.

Hermes torpedoed

H.M.S. Hermes was the first Royal Navy vessel to be fitted with a flight-deck. But she saw little action. She was torpedoed and sunk in October, 1914 – the first year of the war.

In the same year, the liner *Campania* was fitted with a flight-deck and also became the first aircraft-carrier to use a lift in order to bring aircraft up from hangars below-decks.

H.M.S. Argus is usually accepted as being the first true aircraft-carrier. She was built on the hull of an incomplete liner and launched in 1917.

The first ships to be designed and built solely as aircraft-carriers were the Japanese *Hosho*, completed in 1922, and the 11,000-ton *H.M.S.Hermes*, named after her short-lived predecessor, and completed in 1923.

The tank

THE TANK HAS ITS origins in the 'armoured fighting vehicle', and that is an age-old weapon. When Alexander the Great marched into India in the fourth century B.C., he was challenged by King Porus, whose army included 200 fighting elephants, which were almost certainly protected by armour of some kind.

Leonardo da Vinci designed a turtle-shaped 'armoured car' in the fifteenth century A.D., but nothing with which to power it. And the 'shrimps' used in medieval battles – wheeled carts on which cannon were mounted behind a loop-holed armoured shield – were clearly armoured fighting vehicles of a kind.

The invention of the steam engine provided a possible source of power for armoured vehicles. In 1855, during the Crimean War, an armoured assault car driven by steam was built. It had cannon fixed to fire through loop-holes in its sides and – a really 'ancient' touch – scythe-blades fitted to its wheels. But it was never used in action, partly because many British officers considered it 'too terrible a weapon'.

Two Maxim guns

The internal combustion engine, developed thirty years later, offered increased possibilities. Armoured tractors, powered by petrol engines as well as by steam, were used to tow guns in the Boer War. At the same time, an English inventor, F. R. Simms, built an armoured fighting vehicle that consisted of a four-wheeled motor-cycle covered by an armour shield. The driver sat under the shield, which was pierced for one or two Maxim guns, and steered with the help of a periscope. But the machine was too heavy for even the best engines of the time to propel efficiently.

As engines improved, the potential value of armoured cars in warfare increased. In 1914, among the first armoured cars to go into action were built by Rolls Royce. But the Great War saw a more important development – that of the tank.

Churchill insists

As the Great War entered its second year it was clear that some new weapon was needed to break the stalemate of trench warfare: infantry attacks on enemy trenches defended by barbed wire and machine-guns were proving too costly.

Winston Churchill, then First Lord of the Admiralty, had noticed the successes of British armoured cars in the early fighting. He ordered trials to be made with heavier vehicles, which would be able to cross rougher ground. But experiments with armoured tractors failed when the tractors became bogged down in soft patches, and in any case the Army High Command were still convinced that cavalrymen, mounted on 'well-bred horses' could win the battles. They showed little interest in Churchill's 'land battleships'.

Churchill was not put off. He set up a 'Landships Committee' and put naval designers to work. His original project – a 300-ton vehicle, 100 feet long and 80 feet

wide, carrying 12-inch naval guns – proved far too ambitious, but experiments with smaller vehicles proved hopeful and soon even the Army was anxious to share in the scheme.

'Little Willie'

In September, 1915, the first tank (the name was deliberately a misleading one, to fool enemy spies) underwent trials. This was 'Little Willie', powered by a 105 horsepower Daimler engine and capable of two miles an hour on its nine-foot tracks. In order to make its boiler-plate body stable, two large 'tail-wheels' were dragged behind.

Crew of eight

A few weeks later, an improved model appeared. 'Mother' – earlier called 'Centipede' or 'Big Willie' – was 33 feet long, weighed about 28 tons and had a top speed of almost four miles an hour. She carried a crew of eight and was armed with two machine-guns and two six-pounder cannon in 'sponsons' – bulging armoured 'pockets' on either side of her hull. With a few modifications, 'Mother' became the Mark I Tank, and 150 were ordered for the Army in 1916.

First action

The tanks saw action for the first time on September 15th, 1916. A German eyewitness reported:

'*When the German outposts crept out of their dug-outs in the mist of the morning . . . their blood was chilled in their veins. Mysterious monsters were crawling towards them over the craters . . . a supernatural force seemed to impel them on. Someone in the trenches said, "The Devil is coming!" Suddenly, tongues of flame leapt out of their armoured sides.*'

36 tanks attack

Thirty-six tanks lined up for the attack. Just after six o'clock in the morning Tank DI, commanded by Captain Mortimore, bore down on a German position with its cannon roaring and its machine guns crackling. The Germans – not surprisingly – took one look and ran as if the Devil really was after them. All along the Flers-Courcellette sector of the Somme battlefield the effect was much the same that morning. Overhead, a British airman watched and summed up the battle in a single vivid sentence:

'*A tank is walking up the High Street of Flers with the whole British army cheering behind!*'

The atomic bomb

'Sixteen hours ago, an American airplane dropped one bomb on Hiroshima. That single bomb had more power than twenty thousand tons of explosive. It is an atomic bomb. It is a harnessing of the basic power of the universe'.

That was President Truman's statement to the United States on August 7th, 1945.

At seven o'clock the previous morning, Hiroshima had been an important city, the seventh largest in Japan, a harbour on the Ota river delta from which it took its name, meaning 'broad island'. There were about a quarter of a million civilian inhabitants, as well as a garrison of 150,000 soldiers.

At nine minutes past seven, air-raid warnings sounded. There was no real cause for fear; the city had previously had only two light bombing raids, neither of which had caused much damage. Four American B-29 bombers circled the city, and then appeared to change their minds. They flew away.

Raging wind

At half-past seven, the 'all-clear' sirens sounded and the people of Hiroshima left their air-raid shelters. Seconds later, a glaring pinkish-white light filled the sky. It was followed by a blast of searing heat and a raging wind.

The Americans had dropped their 'single bomb', from a great height, near the centre of the city. The centre of the city ceased to exist. A 'zone of death' spread for about three-quarters of a mile from the point of the explosion. Up to three miles away houses were flattened. Thousands of people were killed instantly by blast or heat. Many more suffered terrible burns. Of the survivors, many died within a few weeks from 'radiation sickness'.

What was left of Hiroshima burned for the rest of the day. At evening the fires went out, because there was nothing left to burn. More than 90,000 people were either dead or seriously wounded.

The first test of the atomic bomb had been made in the New Mexico desert on July 16th, 1945, by a team of scientists under the direction of Robert Oppenheimer. Although it represented the peak of his life's work, Oppenheimer himself was disturbed by the awful power he had unleashed, when he saw the fireball 'brighter than a thousand suns' rise over the desert.

Albert Einstein, one of the greatest scientists of all time, whose work had been of considerable help to scientists involved with the project, is said to have remarked, later, 'If I had known that this was to be, I would have lived my life as a simple locksmith'. (*See also Industrial: Atomic Energy*.)

EA-BORNE TAKE-OFF. *November 1910. Commander Ely takes off from the cruiser U.S.S. Birmingham.*

Part 7

COMMUNICATIONS

Alphabets, printing and Braille, the newspapers and the pen, shorthand, maps and semaphore – they are all methods of communication, some very old, some not so old. Now we also have the radio and television, the record-player and the tape-recorder, the telephone and the satellite, but these are only more sophisticated methods of getting the message across, whether verbal or pictorial, from one person to another.
Despite the new ways of communication the old ones have not disappeared: we make phone calls but we still write letters.
We use some or all of these methods almost every day – we consult a map, listen to some music, watch a programme or maybe we read a book or listen to someone talking to us with words that we understand.
Perhaps, one day, all these methods may give way to what one supposes would be the perfect means of communication – telepathy. No wires, no instruments, no paper! In which case, we will try and track down the first official telepathist, though of course people have been claiming to read the minds of others for centuries . . .

THE FIRST PHOTO OF A PHOTO BEING TAKEN. *In 1839, the English scientist, William Fox Talbot, described his now famous invention – the first negative-positive process in photography. In the photographic world, he also scored this notable first – the first photograph of a photograph actually being taken.*

A written language

THE FIRST WRITTEN LANGUAGE was probably Sumerian, though no one would use it now for conversation or communication. The language was discovered on some clay tablets which were inscribed with small pictures representing words and phrases, found at Warka, in Iraq. Experts have estimated that the tablets were inscribed in about 3300 B.C.

Some archaeologists maintain that tablets found in Iran are even older. These were inscribed with a form of Elamite language and are said to date from at least 200 years earlier. But it has not yet been established satisfactorily that the Elamite tablets are as old as they are claimed to be.

A name

THE EARLIEST WRITTEN RECORD of a person's name is as old as the earliest written language. The clay tablet on which the name of a Sumerian citizen was inscribed was found at Jamdat Nasr, 40 miles from Baghdad. It dates from about 3300 B.C.

Translated into today's alphabet the name of the citizen was *En-lil-ti*. He was probably a priest.

An alphabet

ALPHABETS ORIGINATED in simple sketches of animals and objects, called ideograms, which consisted mainly of straight and curved lines, circles, squares and triangles. During the course of centuries these signs became more sophisticated, until they developed into the Ancient Egyptian hieroglyphics. Some of these, chiselled on stone nearly 5,000 years ago, still exist.

In time, however, picture and symbol writing became so complicated that few people were able either to read or write the various forms. New sets of symbols were invented that could be joined together to make up any word no matter how long it might be, until eventually these sets of symbols became the alphabets we know today.

Semitic tribe

The world's first alphabet, as distinct from picture-writing, was evolved about 4,000 years ago by a Semitic tribe living in Sinai. One of the letters used in the alphabet was 'O', which has remained unchanged – so you might say that 'O' is the oldest of all the letters in the 66 alphabets in use throughout the world at this time.

The word 'alphabet' is derived from the first two letters of the Greek alphabet: *alpha beta*.

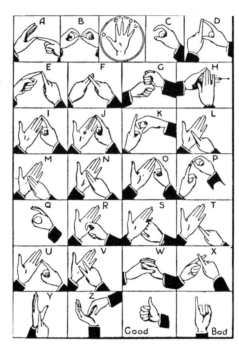

ABOVE: SPEAKING BY SIGNS. *Charles Epee invented the first system of "speaking" by means of signs. His method is the basis of all manual alphabets in use today.*

A manual alphabet

IN 1755 CHARLES EPEE opened a school in Paris to teach deaf children to speak by means of signs with the hands and fingers. His system, which he had invented two years earlier, used various positions of both hands and all the fingers and thumbs to represent the letters of the alphabet and the nine numerals.

Epee's manual alphabet was the basis for all the deaf and dumb sign alphabets in use today. At one time, all deaf people used the manual alphabet; nowadays most with any sense of hearing, however slight, rely on electronic hearing aids, others use lip-reading, even though quite deaf.

Braille

THE FIRST SUCCESSFUL reading system for the blind was invented by Louis Braille.

Braille was blinded in 1812, when he was three years old. He was playing with a heavy needle in his father's shoemaking workshop when his hand slipped and the point of the needle entered the corner of his eye. The injury did not seem very serious at first but some dirt also got into

BELOW: READING BY DOTS. *A French artillery officer invented the raised-dot alphabet for the blind – but it was his blind pupil, Louis Braille, who made it workable.*

BRAILLE ALPHABET

A	B	C	D	E	F	G	H	I	J

K	L	M	N	O	P	Q	R	S	T

U	V	X	Y	Z	and	for	of	the	with

W	Oblique stroke	Numeral sign	Poetry sign	Apostrophe sign	Hyphen	Dash

Lower signs	,	;	:	.	!	( )	?	" "	; ...

the eye from the leather. Gradually, as infection spread also to his other eye, his sight began to fail and within a couple of years he was completely blind.

When he was 13 he was sent to the National Institute for the Young Blind in Paris. At that time the blind were taught to read by means of thick cloth letters of the alphabet stuck on paper to form words. As you can imagine, this was a pretty slow and heavy process: not only did it take a long time to stick the letters on the paper but the completed sheets were extremely thick and took up a lot of space.

"Night writing"

A French artillery officer called Charles de la Serre had recently visited the Institute and had been allowed by the Principal to test the pupils with a system he had invented of dots raised on sheets of cardboard to represent the letters of the alphabet. Originally Serre had designed the dot system in order to help his soldiers read written orders in the dark: he called it 'Night Writing'.

Louis Braille liked the idea but thought it too complicated: twelve dots were needed for each letter. So he simplified the system down to a maximum of six dots, in various combinations and numbers, to represent the letters of the alphabet, punctuation signs, and mathematical numbers and symbols.

No encouragement

The first book in Braille appeared in 1834, but Louis received no encouragement and even the Institute forbade the teaching of the new system.

Ill from tuberculosis, but undiscouraged, Louis wrote many articles and pamphlets explaining the system, but it was not until ten years after his death that Braille came into general use with the blind. Later, several improvements were made to Louis Braille's system to speed up reading and writing.

Newspapers

THERE WERE PROFESSIONAL NEWS-GATHERERS in Rome as far back as the Fifth Century before Christ. For a fee they would send information concerning important events to clients who lived too far from the capital to keep in close touch themselves. But the first idea for a newspaper, as we understand the word, was thought up by Julius Caesar.

When Caesar became Consul in 60 B.C. he started the *Acta Diurna*, a daily bulletin posted in the Forum and devoted to government announcements.

● The first newspaper to be printed regularly was the *Avisa Relation oder Zeitung*, which began publication in Augsburg, Germany, in 1690.

● The first English newspaper to appear regularly was the *Corante, or Weekly Newes from Italy, Germany, Hungarie, Spaine and France*. The first issue of the *Corante*, which was only a single sheet, was in September, 1621. *Corante* means 'current' or 'up-to-date'.

● In November, 1641, the first newspaper devoted to events in England appeared. This was the weekly *Diurnal Occurrence* (Daily Occurrences).

● The first illustrated newspaper was published in London. It was called the *Mercurius Civicus* (Civic Mercury) and was illustrated by woodcuts.

● The first daily newspaper was the *Daily Courant*, published in London, in 1702.

Shorthand

SHORTHAND – phonetic or sound writing – was not invented, as some people seem to believe, by Sir Isaac Pitman, who designed today's most popular system. There have been variations of shorthand in use for thousands of years.

The earliest accepted system of shorthand was probably that invented in 63 B.C. by Marcus Tullius Tiro, secretary to the great Roman statesman and orator Cicero. When Cicero wished to keep a record of his speeches in the Roman Senate, Tiro devised a system of brief signs which could be written rapidly enough for the listener to be able to take down a speech word for word as it was spoken. Tiro used the system for the first time to record Cicero's speech against Catiline.

Tiro's shorthand was later used extensively by the Romans not only for keeping records of speeches but also for recording trials.

A printed book

THE OLDEST KNOWN PRINTED BOOK is a collection of poems which was produced in China in 868 A.D. The Chinese first started carving letters and pictures in relief on blocks of wood nearly two thousand years ago. These blocks were then inked and stamped on single sheets of paper. But it took the Chinese several centuries to discover how to bind the sheets together to form a book.

Movable type

MOVABLE TYPE uses separate letters and words – instead of blocks – which can be moved about on the line of type. In 1456 Johann Gutenberg of Mainz used movable type for the first time in the West to print an edition of the Bible. This became known as the Gutenberg or Mazarin Bible.

Although Chinese methods of printing had never reached the West, German craftsmen had discovered for themselves the use of carved blocks of wood for printing in 1400, and this was the method they used until 1456.

Gutenberg later invented moulds from which individual letters of the alphabet could be cast. He also invented an ink which could be transferred to paper from metal type.

A type-setting machine

OTTO MERGENTHALER – a German watchmaker who became a naturalised American citizen – built his first type-setting machine in 1885, after paying a visit to a printing works and being appalled by the amount of work involved in setting the type. Before he produced his time-saving machine every piece of printed matter had to be set slowly and laboriously by hand.

The operator of Mergenthaler's typesetter tapped out the words on a keyboard similar to that of a typewriter. As the words were tapped out each key released a piece of metal into which was moulded the letter of the alphabet as was needed. The pieces of metal were then brought together to form the necessary words.

Mould of words

When there were enough words to form a line of type in the pages of the newspaper or book that was being set, the row of metal pieces was carried by the fingers of a mechanical arm and placed against a pot of molten metal which provided a mould of the words. The metal cooled and hardened very quickly and the line of type – called a 'slug' – now ready for printing dropped into a tray.

Because the machine set the words in a line at a time, Mergenthaler called it a 'linotype' (line o' type). The linotype was first put to practical use on July 3rd, 1886.

The following year an American called Tolbert Lanston invented a machine that cast the letters singly. This overcame one of the great disadvantages of the linotype, that if the operator made a mistake it was necessary to reset the whole line of type in order to make the correction. Any corrections on Lanston's 'monotype' (setting the letters one by one) could be made simply by changing a single right letter for the wrong one.

The pen

THE FIRST REAL pen was invented by the Ancient Egyptians. It was made of a piece of copper cut in the shape of a nib and fastened to a reed. Later the copper nib was replaced by a hollow reed with the end cut to form a nib. The pen was dipped into an ink made from a mixture of ground charcoal and gum. Many papyrus manuscripts were written with this kind of pen and ink.

The quill pen was invented when paper was introduced about 1,000 years ago, because earlier writing instruments were unsuitable for writing on paper. The tail or wing feather of either a goose or swan was pointed at the end and split down the centre. In fact the name 'pen' comes from the Latin *penna*, which means 'feather'.

The first steel pen-nib was made in London in 1803 by John Wise. As the nib had to be cut and shaped by hand it was naturally expensive and as a result not a commercial success.

Seventeen years later, however, three English engineers, Gillot, Mitchell and Perry, formed a partnership and invented a machine to make an improved form of steel pen. This could be sold for a few pence a dozen – a good deal cheaper than John Wise's.

Ink gushed out

It was the fountain pen that was the most revolutionary of all writing instruments – a pen that did not need to be dipped constantly in ink but carried its own supply. Since the middle of the nineteenth century there had been numerous attempts to produce a fountain pen, attempts which had all failed, either because the ink was not allowed to flow freely enough or because it gushed out too freely and spread all over the paper.

The first practical fountain pen was patented in 1884 by a man with a name still famous for fountain pens today – Lewis Waterman.

Waterman was an insurance agent in New York, accustomed to carrying with him a bottle of ink and a steel-nibbed pen with which his customers could sign forms. But the forms were too often ruined by the bottle being upset or the pen smudging. So Waterman, a keen amateur mechanic himself, set to work to design a fountain pen that could be trusted. The main problem – the problem of the ink-flow – he solved by cutting a small groove on either side of the component supporting the nib: air flowed up one channel and ink down the other. The secret was that the intake of air up one side controlled the downward flow of the ink. And the nib was mounted in such a way that ink could flow to its tip only when the nib was pressed on the paper.

Signed guarantee

Waterman's confidence in this invention was so great that he gave up his insurance business and began manufacturing fountain pens. With each one he gave a signed guarantee against any defects.

The pencil

THE GREEKS probably invented the first pencil, or non-ink writing instrument, about 3,500 years ago. This was a piece of bone or ivory with a sharp point, which was used for writing on wooden or metal tablets coated with wax. The tablets were meant only for taking temporary notes: the writing could be rubbed out simply by smoothing the wax, which could then be used again.

So-called 'lead' pencils do not contain lead at all, but a mixture of clay and graphite which has been baked hard. The first attempt to make pencils of this kind was in the middle of the fourteenth century when graphite was discovered in a mine in Cumberland. The mineworkers mixed graphite with clay and rolled the mixture into sticks – a crude pencil without any covering.

The first wood-covered pencil – as we know the pencil today – was made by a Frenchman called Conté, in 1795. He shaped the graphite and clay mixture into thin strips, baked them and then inserted them into hollowed-out cylinders of cedar wood. A covering of thin glue kept the graphite mixture from slipping out of its wooden case.

LEFT: THE FIRST MACHINE TO TYPE OUT... TYPE. *In the 1880's, all type was still set laboriously by hand, letter by letter – until a watchmaker invented the "linotype" machine.*

Printed for & Sold by BOWLES & CARVER.

THE *ORIGINAL* BATH MAIL COACH.
Invented by Mr Palmer *of* BATH, *and Supported by* GOVERNMENT.

Nº 69 Sᵗ Pauls Chur

Mail coaches

THE CARRIAGE OF MAIL was revolutionised by the introduction of a mail-coach service in 1784. The coaches were specially built to travel fast and were drawn by teams of first-class horses at a speed of between nine and ten miles an hour. The horses were changed every ten miles.

The first mail coach to go into service was between London and Bristol, via Bath. The pioneer venture was so successful that within two years the new mail coaches were travelling on all the principal roads out of London and, within 50 years, 28 mail coaches a day were leaving the General Post Office in London for destinations throughout the United Kingdom.

1784 ... THE FIRST OF THE MAIL COACHES. *First-class horses, changed every ten miles, this was the first mail coach. It ran between London and Bristol. Speed ... ten miles an hour.*

Before the introduction of mail coaches in Britain, the mail was carried by mounted post-boys who were scheduled to travel between the main towns and cities at a speed of seven miles an hour in summer and five in winter. However, the roads were so bad that the post-boys seldom went faster than two miles an hour. So the mail was rarely on time!

Even by the middle of the eighteenth century, when the roads gradually improved and most of the mail was carried by stage-coaches and stage-waggons, the regulation speed was seldom reached.

FEED CLOSED   FEED OPEN

THE "PENOGRAPHIC" FOUNTAIN PEN, *patented by John Scheffer in 1819, was one of the first "reservoir" pens. A cock opened and shut the feed-pipe from the reservoir of ink to the nib. A lever, on the side of the pen, could be depressed to exert pressure on the reservoir to force through more ink when necessary.*

Mail trains

LETTERS WERE FIRST CARRIED by train in November, 1830, between Liverpool and Manchester. Within eight years mail trains were operating between London, Liverpool, Manchester and Bristol.

Very soon the mail trains completely replaced the mail coaches: in 1846 the last of the London-based horse-drawn mail coaches left London for Norwich, via Newmarket.

The first travelling post office went into service in 1838. A train on the Grand Junction Railway had a coach attached to it in which letters could be sorted and bagged during the journey.

Mail only

Apparatus on the side of the coach and at the edge of the railway track made it possible for the mail to be picked up and off-loaded while the train was travelling at speed. Within a few years travelling post offices were operating on most of the principal railways out of London.

The first railway to be built solely for the carrying of mail was opened for service in 1927. The trains, which have no drivers, run through a system of tunnels beneath London and link the main-line railway stations with the chief sorting offices. This cuts down the heavy delays previously caused by traffic congestion when the mails were moved through the streets.

1830. THE FIRST LETTERS *to be sent by rail went on this train from Liverpool to Manchester. Sixteen years later, the last horse mail coach was withdrawn.*

The postmark

POSTMARKS WERE FIRST USED on letters in 1680 by William Dockwra, a London merchant who set up a private letter-delivery service in London. All letters and packages passing through his receiving offices were stamped with the date and the name of the office.

When Dockwra's license to operate his post office lapsed, it was not renewed, and postmarks were not used again until 1840, when Rowland Hill's penny post was established. The first postmarks to cancel Hill's stamps were at first simply a coloured cross to obliterate the stamp.

Later the postmarks cancelled the stamp by impressing on it the name of the town or district where the letter had been posted and the date and time when it was received at the post office.

In the First World War the first slogans were added to the cancellation postmark – for instance, slogans encouraging people to buy war loan bonds. Other postmarks have included propaganda for fuel saving, safety first, salvage and national savings. A British postmark including a design of victory bells celebrated the end of the Second World War.

The postage stamp

THE 'PENNY BLACK' – the world's first postage stamp – was stuck on a letter at the General Post Office, London, in May, 1840. It replaced the special envelopes, called Mulready's, which had been sold by the Post Office as proof that postage had been prepaid by the sender.

The Penny Blacks were printed in solid sheets of 240 and, as they were not perforated, had to be cut off with scissors. They had been designed six years earlier by John Chalmers of Dundee in anticipation of the introduction of penny post. Since the issue of the Penny Black, nearly 150,000 different postage stamps have been issued throughout the world, and that number increases by nearly 2,000 new issues every year.

In recognition of her lead, Great Britain is the only country in the world not required under international postal law to show the name of the country on her stamps.

The pillar box

ON NOVEMBER 23RD, 1852, a 'Roadside Letter Box' – that was the name the Post Office gave it – was set in the pavement of David Place, St. Helier, Jersey. It was the world's first pillar box – eight-sided, with a horizontal slit in the top for posting letters and a notice of the times of collection.

WHEN PILLAR BOXES WERE GREEN. *The Channel Islands had the world's first post boxes – they were eight-sided and painted in green. Anthony Trollope, the novelist, thought of the idea. In 1855, pillar boxes were erected in London.*

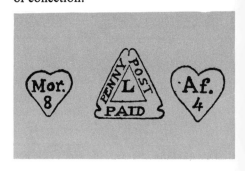

THE FIRST POSTMARKS. *1680. A London merchant opened the first postal service – a private one. The centre stamp, marked "L" for Lambeth, was his receiving stamp.*

How the Christmas Card was born

In the early days of postal services letters had to be taken by the sender to a receiving office for delivery to the addressee. In country districts this could frequently mean a journey of several miles. Later certain shopkeepers were authorised to collect letters from their customers and hand them on to an official post office messenger. In Paris, and other large cities and towns in France, shopkeepers were given a locked box with a slit on top for the letters, so that there could be no prying. The boxes were opened and the letters collected daily by a messenger.

Iron boxes

When the penny post was introduced in 1840, there was a steady increase in the use of the postal service in Britain. The Post Office decided that some more convenient method of collection was necessary. As is often the case, decision did not turn into action for another 11 years.

In November, 1851, Anthony Trollope, a post office official – better known as a novelist to us – was sent to Jersey to survey the postal services there and suggest methods for improving them. His proposal that iron letter boxes should be mounted on the street pavements or fixed to the walls of buildings was, for a change, acted on promptly.

Two or three days after the first pillar box was set up, three more were introduced in St. Helier. In February, the following year, the scheme was extended to the rest of the Channel Islands. One of the original pillar boxes, which was set up in Union Street, St. Peter Port, about 1856, is still in use – the oldest letter box in the world in service.

The experiment in the Channel Islands proved so successful that in 1855 pillar boxes were erected in London. At first there was no standard design: the boxes were of all shapes and sizes; many were highly decorated. It was not until 26 years after they were first introduced that they were standardised to the now familiar cylindrical shape. Originally painted green, the present red was first used four years before the standard shape was established.

Post cards

THE FIRST picture postcards had a view of Lake Garda, in Italy. They were printed and sold in 1865 by Cesare Bertana, an Italian stationer. They quickly became popular with the tourists, who enclosed them in their letters for home.

During the Franco-Prussian war of 1870 picture postcards of local views were sold in large numbers to German troops as souvenirs of the places through which they marched.

Eddystone lighthouse

The first British picture postcard was printed in 1891. It showed a view of Eddystone lighthouse and was issued in conjunction with a Royal Navy exhibition.

The first postcards officially sold by a post office were put on sale in London in 1870. These had no picture: one side of the card was for correspondence, the other for the address. The card cost three farthings, including the postage stamp printed on it.

Christmas cards

'I hate those redbreasts!' declared *Punch* in 1869, fed up with Christmas Cards already, though they had only been marketed in a big way by the firm of Marcus Ward & Co., of London and Belfast, within the previous nine years.

The earliest Christmas Card had a much more serious theme, though it was attacked by the Temperance Movement for showing a picture of a family eating and drinking. That was only part of the card that J.C. Horsley designed for Henry Cole in 1843. The card was in the form of a triptych with a medieval theme. Orna-

THE FIRST CHRISTMAS CARD *was designed for Sir Henry Cole's art shop in 1843. Actually, he was hoping to sell these cards in order to improve the public's appreciation of art. At that time, however the Industrial Revolution had broken up many English families so that people, for the first time, were not at home with their relatives at Christmas time. The cards were used to keep the link.*

ments divided the sections and were derived from Durer's Prayer Book of Maximilian. The central section was the one that brought on the criticism of the Temperance people; the side panels showed good works – clothing the naked and feeding the hungry.

Scattered families

The card was not intended for Cole's personal use. It was to be sold at his art shop in Old Bond Street in an attempt to raise public taste. As Horsley said of Cole: 'He devoted much of his time in getting artistic treatment applied to unconsidered trifles as well as to the weightiest matters.' In fact, Sir Henry Cole had a particular interest in industrial and applied art and later became the first Director of the Victoria and Albert Museum.

One reason for the success of the Christmas Card in the 1860s was that the Industrial Revolution scattered families and weakened or destroyed totally the tradition of celebrating Christmas as a family, village or town festival. The cards helped to maintain the links with those who had moved away; usually, therefore, they showed scenes from their old world.

Air mail

AERIAL POST began in Britain in June, 1911, when a bag of letters was loaded onto an aeroplane at Hendon aerodrome, London, and flown to Windsor. The letters had been stamped and were passed through the post office; they formed part of the coronation celebrations for George V. However these were not, as has often been thought, the world's first airmail: there was a regular aerial post in the Franco-German war of 1870-71. Whilst Paris was under siege, balloons were used to carry provisions into the city and help those who wanted to escape. The balloons also carried a systematic postal service.

Milestones

MILESTONES WERE FIRST set up in a regular fashion by the Romans under Julius Caesar. These milestones were round stone pillars six feet high placed 1,000 'passus' apart – equal to 1,617 yards. The stones marked the distances from Rome along all the highways of the Empire.

In the Forum at Rome there was a bronze-gilt milestone on which were carved the names and distances of the chief towns and cities served by the roads which led out of Rome's thirty-seven gates.

The first official milestones in Britain were set up in 1593.

A map

THE FIRST MAP of which we have any record is a town plan of Babylon. This was drawn on a clay tablet about 4,300 years ago. It is now in the British Museum.

The first map to use lines of longitude and latitude was drawn in about 150 A.D. by an Egyptian scientist called Ptolemy, who was the first geographer to realise that the world was round.

Ptolemy's map only showed part of the world because only part of the world was then known to him. Alexandria, which was where he lived, stood at the centre of the map, and the position of every important place as well as its distance from Alexandria was marked on the map.

Treasure chart

The earliest treasure chart that has survived was drawn about 1,320 years before Christ. It appears in a papyrus scroll called the Turin Papyrus and describes the way to the entrance of an Egyptian gold mine. The chart also gives a sketch of the mine workings.

THE LIGHTHOUSE OF ALEXANDRIA. *The world's first lighthouse, the Pharos of Alexandria. In 280 B.C., Ptolemy II, of Egypt, built it on the island of Pharos, at the entrance to Alexandria.*

A lighthouse

DURING THE EARLY DAYS of seafaring, sailors were often warned of dangerous rocks and other hazards – sunken ships for instance – by bonfires lit on the cliffs or high ground along the coast.

No permanent lighthouses, so far as we know, were built until 280 B.C. when Ptolemy II of Egypt ordered a tower to be built on the island of Pharos at the entrance to Alexandria harbour. The tower, called the Pharos, was made of stone, 500 feet high, and a fire was kept going constantly on top of the tower to guide sailors into the port.

The Pharos was considered to be one of the Seven Wonders of the World. It remained in operation until about 400 A.D., when it was partly destroyed by an earthquake – after nearly 700 years service! It was finally levelled to the

ground in 1375 by a second earthquake. Many of the original stones were used in the rebuilding of Alexandria harbour.

1732... THE FIRST LIGHTSHIP. *Oil lanterns hung from the top spar of this old ship, anchored in the treacherous waters of the Nore, in the Thames estuary.*

ABOVE: DESTINATION . . . WINDSOR CASTLE. *As part of George V's coronation celebrations in 1911, letters were loaded on to this aeroplane at London – and flown to Windsor Castle. This was Britain's first post by aeroplane.*

RIGHT: LOADING THE PLANE. *The idea of sending letters by plane was certainly revolutionary. It was also a method which clearly demanded strong nerve . . . if we study the plane.*

A lightship

INSTEAD OF A LIGHTHOUSE – where it might not be possible to build one or where the danger is only temporary – reefs, shoals and other underwater obstructions are often marked by lightships, anchored over the danger area.

The first lightship was an old sailing vessel which was anchored in the Nore, in the Thames estuary, in 1732. Its light came from an oil lantern hung on the top spar of the main mast.

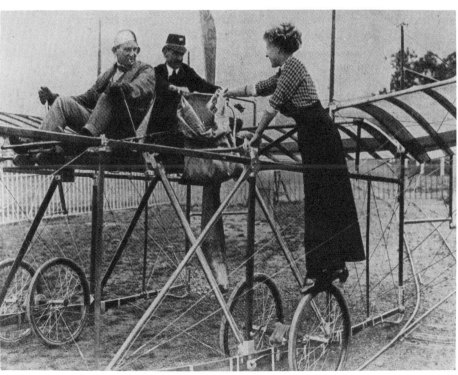

Semaphore

MEN MUST HAVE passed messages to each other over long distances by waving their arms from the very earliest times but no one thought of working out a proper system of arm signalling to spell out words according to an alphabet until 1767, when John Edgeworth invented the first semaphore, as we know it.

The name 'semaphore' comes from two Greek words: *sema*, meaning a 'sign', and *phora*, meaning 'to carry'.

Edgeworth was a racehorse owner who lived near Newmarket. He was accustomed to attending all the races in which his horses took part but in 1767 he was taken ill and was unable to leave his house. Naturally he became extremely frustrated in his anxiety to hear how his horses had got on. So he had installed at the racecourse a tall pole on which two wooden arms were hinged. He could see the pole from his house.

Ropes and pulleys

By means of ropes and pulleys the arms could be moved into various positions to indicate the letters of the alphabet. One of his grooms was taught how to operate the semaphore, and Edgeworth's illness became a little more bearable.

His idea was not extended until the Napoleonic wars, when the French government set up chains of semaphore relay-stations to keep Paris in touch with the armies on her frontiers. Soon afterwards the British Admiralty set up semaphore stations in a line, each one eight miles from the next, to communicate orders from London to the naval bases in Portsmouth and Chatham. The name 'Telegraph Hill', which appears on maps, reminds us where some of the relay stations stood.

The telescope

IN 1608 the Dutch government decided that it was necessary to make a more thorough study of the stars in order to assist navigators. A prize was offered for a reliable telescope. On October 2nd, the prize was won by Johannes Lippershey, a Dutch spectacle maker, for a refracting telescope which he himself made.

Scientists of Ancient Greece and Egypt had understood the magnifying power of a lens but no serious study of the subject had been made until the thirteenth century, when Roger Bacon, a Franciscan monk, wrote a manuscript suggesting that lenses could be used to make distant objects seem nearer. Bacon never put his theories into practice.

Galileo's telescope, which he built the year after Lippershey won the prize, was a slight improvement. It was also the first to be used solely for astronomical purposes. Within a short time of completing it, Galileo had discovered four of the moons which circle Jupiter, the rings of Saturn, the mountains and plains of the moon, and spots on the sun.

Photography

THE BASIC PRINCIPLES of photography were discovered long before the camera, as we know it, was invented.

During the tenth century A.D. Arab scientists discovered that by allowing light to enter through a small hole in the wall of a darkened room they could produce an upside-down image of the scene outside the room on the wall opposite the hole. Arab astronomers used this method to watch eclipses of the sun without risking damage to their eyes by looking directly at the light.

Nothing more was heard, however, of the Arab idea until the fifteenth century, when an Italian made a large wooden box and painted it black on three of its inner sides. On one of the darkened sides he drilled a pinhole, opposite to the unpainted, white side. The light entered

through the pinhole and cast a picture of the scene outside the box on the white surface. Later a lens was fitted over the pinhole to give a much sharper picture.

The Italians called their picture box a *camera obscura*.

In the eighteenth century a German artist made a small and efficient *camera obscura*, about the size of a modern box camera, which he used as a guide when painting scenery. Unfortunately there was no way of making the picture permanent: once the light had gone, so did the image.

The first successful use of light to obtain a permanent picture was made by Thomas Wedgwood, son of the famous potter Josiah Wedgwood. He made his first 'photograph' in 1802 by laying fern leaves on a sheet of paper which had been soaked in a solution of silver nitrate. When the ferns and the paper were exposed to sunlight, the paper surrounding the leaves turned black while the paper

DAGUERREOTYPE PORTRAIT. *The metal portrait produced by the Daguerreotype camera. They were very desirable between 1837 and 1867 and this process really established photographic portraiture.*

under the leaves, which received no light, remained white.

Piece of pewter

However the degree of 'permanency' was only limited: the fern pictures did not last unless they were kept in the dark, where it was pretty impossible to see them. Immediately the fern leaves were taken off the paper in the light, then the paper beneath the leaves also became black!

The next step, 25 years later, did at last result in what we can probably consider to be the first real photograph.

Joseph-Nicéphore Niépce, a retired army officer, made a simple type of box camera fitted with a lens in 1827. Inside the box he placed a square piece of pewter coated with a substance called bitumen of Judea. When the photograph was taken he washed the plate in oil of lavender, which dissolved all the bitumen that had

not been exposed to the light, leaving the outline of the scene that had been photographed.

Again there were difficulties: Niépce's photograph could only be seen when the plate was held at a right angle, because there were reflecting differences between the coated and the uncoated areas. Anyway, Niépce's photography, although a

ABOVE: THE BEGINNING OF THE "BOX" CAMERA. *In the 10th century, Arab scientists invented the "camera". In the 15th century. Italian scientists took over the idea and built "box" cameras, as shown above, called* **camera obscura.** *At this time, there was no way of making the image permanent.*

BELOW: THE FOX TALBOT CAMERAS. *These were the first cameras used by the English pioneer of photography, William Fox Talbot.*

THE START OF MODERN PHOTOGRAPHY.
William Fox Talbot invented the paper negative method. This picture of Lacock Abbey, Wiltshire, England was printed from one of Talbot's first negatives – made in 1841.

step in the right direction, was not very practicable: an exposure took eight hours and the photographs themselves were indistinct.

Before he died, however, Niépce entered into a scientific partnership with Louis Daguerre, a French scene-painter, to carry out photographic experiments. Four years after Niépce's death in 1833, the first *daguerreotype* appeared. Basically it was an improvement of Niépce's method. But the exposure only took 30 minutes and Daguerre succeeded in 'fixing' his photographs so that they would not fade – probably his greatest achievement.

Daguerreotypes became a tremendous commercial success, particularly for portraits. They remained in use for almost 30 years.

The first book to be illustrated by photographs was produced in 1844 by an Englishman – William Fox Talbot. It was called *The Pencil of Nature*, and all the photographs were taken by Talbot himself. His big contribution to the development of the camera was to introduce strips of treated paper instead of metal plates; also, any number of positives could be printed from his negatives. Moreover – and perhaps for the first time – Talbot's photographs were really *accurate* pictures of the objects or scenes photographed!

Three plates

Seventeen years later the English scientist James Clerk Maxwell took the first photograph in colour – a piece of tartan ribbon. The ribbon was photographed three times on separate plates, each of which was sensitive to certain colours. The three plates were subsequently superimposed in a magic lantern and projected on to a screen.

●THE FIRST ROLL FILM was invented by George Eastman and William Walker in 1885.

●THE FIRST FOLDING CAMERA for amateurs was put on sale in 1890.

●THE FIRST CAMERA WITH ROLL FILMS that could be loaded into it in daylight was introduced in 1898.

●THE FIRST BOX 'BROWNIE' camera was put on sale in 1900. It cost one dollar and used film costing 15 cents a roll.

●THE FIRST COLOUR FILM for amateur photographers was put on the market by the Eastman company. It was invented by two musicians, Leopold Godowsky and

Leopold Mannes, and it was called Kodachrome.

Cinematography

CINEMATOGRAPHY is the technical name for moving pictures. The word comes from two Greek words: *kinema*, meaning 'movement', and *graphein*, meaning 'to draw'.

Two major problems had to be solved before the moving picture became possible: a camera able to take photographs in rapid succession, and some material more flexible than glass on which to take the photographs. Both problems were solved by William Friese-Green, a professional photographer with a studio in Piccadilly, London.

Friese-Green hit on the idea of using a thin strip of celluloid suitably treated with chemicals on which to take the photographs. Down each side of the film were evenly-spaced holes. The camera which he designed had a handle at the side which when turned revolved two small cog wheels mounted on either side of the lens.

As the film was wound behind the lens the shutter opened and closed several times a second to take a succession of still photographs. When the photographs were projected on to a screen very rapidly, one after the other, they gave the impression that the objects which had been photographed were actually moving.

A little jerky

The photographer took his first film at Hyde Park Corner on a spring morning in 1890. That evening he developed and printed the film and ran it through his projector. The film was certainly a little jerky but it did show cabs and pedestrians moving about.

In his excitement, Friese-Green ran into the street and persuaded the first person he met – a policeman on the beat – to come in to see his moving picture!

One earlier experiment is worth mentioning: the first man to use photographs to give an illusion of movement was Edward Muybridge, an English photographer on a visit to America.

In 1872, he was commissioned by the Governor of California to photograph race-horses in action. To do this, Muybridge laid a battery of cameras side by side along the edge of the racetrack. Each camera shutter was connected by a strong spring to a thread stretched across the track. As the horse galloped along the track it broke each thread in succession and so activated the shutter of each camera as it passed.

Actual movement

Since the photographs were taken on glass plates, there was no way of running them continuously through a projector. Therefore they had to be looked at separately. All the same, the series of photographs provided a lot of new information about the actual movement of horses and perhaps laid the foundation for the cinema, which the ingenuity of Friese-Green developed in his film.

● THE FIRST PUBLIC SHOWING of a film was on December 28th, 1895, in the Indian Salon of the Hotel Scribe in Paris.
● THE FIRST CONTINUOUS performance cinema was opened in Los Angeles, California, on April 2nd, 1902.
● THE FIRST FILM with the music synchronised, or in time with the action, was *Don Juan*, released on August 26th, 1926.
● THE FIRST FILM in which the speech of the characters was synchronised with the action by a sound track on the film was the 'talkie', *The Jazz Singer*, released on October 26th, 1927.
● THE FIRST ALL-TALKING film was *Lights of New York*, shown in New York on July 6th, 1928.
● THE FIRST FILM IN COLOUR was a newsreel of the Delhi Durbar, 1912. Each photograph on the reel of film was coloured by hand.
● THE FIRST SUCCESSFUL film photographed in natural colour was *Becky Sharp*, released in 1935.
● THE FIRST CAMERA, projector and film for home movies was introduced by George Eastman in 1923.
● THE FIRST COLOUR FILM for amateur home movies was introduced in 1935.

● THE FIRST CINEMA opened to the public was the Electric Theatre. It formed part of a circus in Los Angeles, California, and showed its first film on April 2nd, 1902. The first building properly designed as a cinema was the Biograph Cinema in Wilton Road, London. There were seats for 500 people and the cinema opened in 1905.

Television

ON THE EVENING OF January 27th, 1926, a number of scientists, including members of the Royal Institution, gathered in a laboratory in an upper room of a building in Frith Street, London, to watch the first demonstration of a television.

The television itself was little more than a collection of odds and ends picked up at junk shops: a large cardboard disc with bits of glass around it, behind which were several old electric motors and a mass of tubes and pieces from old radio receivers.

The man who had put it all together was John Logie Baird – slim, pale-faced, in his late thirties, nervously turning the knobs on a small control panel. Seated on a chair in front of the cardboard disc was a sixteen-year-old actor – the world's first television actor, you could say.

Baird focussed

As the actor moved his head from side to side, Baird focussed and tuned his transmitter until, on a receiver in the same room, the audience saw a picture of the actor speaking and moving. The demonstration was repeated with a receiver in an adjourning room.

To be honest, the image on the receivers was blurred and faint, but Baird's 'Televisor' (as he called it) did show for the first time that it was possible to transmit and reproduce real "live" scenes instantly.

There had been other pioneers, but they had never before achieved what Baird did. In 1923, for instance, Charles Jenkins sent a silhouette, or shadow picture, of President Warren Harding by radio from Washington to Philadelphia – a distance of 130 miles. And during the next couple of years other American inventors carried out similar experiments, one of which was the successful transmission of a photograph. But all these transmissions were of still pictures only.

Outdoor pictures

In June, 1928, Baird took the first outdoor television pictures.

On August 22nd, in the same year, engineers of the American General Electric Company carried out the first picture news broadcast – a televising of Governor Alfred Smith of New York accepting the Democratic nomination for the Presidency.

The first television play to be broadcast was the *Queen's Messenger*, transmitted experimentally from the General Electric Company's laboratories in Schenectady, New York, on September 11th, 1928.

The world's first public television broadcasting service began at Alexandra Palace, London, on November 2nd, 1936.

A submarine cable

THE WORLD'S FIRST successful submarine cable was laid in 1850 between England and France. Messages were sent regularly between the two countries until the autumn of 1850, when the cable was cut accidentally by a French trawler. A new cable was laid along the floor of the Channel the following year, since when England and France have remained in more or less constant telegraph and telephone communication.

As early as 1795 an officer of the Royal Engineers sent an impulse of electric current through a cable laid on the bed of the River Medway, near Chatham. When a switch at one end of the wire was closed, current passed through the wire and moved a galvanometer at the other end. A galvanometer is an instrument that detects an electric current.

Unfortunately water leaked through the tarred hemp which was wound round the wire to act as an insulator. Anyway, there was at the time no known way of

1858. THE FIRST OCEAN CABLE. *The HMS "Agamemnon", lays the first telegraph cable between Britain and America. Queen Victoria and President Buchanan exchanged greetings.*

varying the current to spell out a message: Cooke and Wheatstone were not to develop the electric telegraph for another 45 years.

The first Transatlantic cable was laid in 1858, and in August of that year the first messages were telegraphed between America and Britain – short greetings between Queen Victoria and President Buchanan. Two months later an engineer sent an accidental surge of high current through the cable: the insulation broke down and the cable went dead.

No further attempt was made to lay a second Transatlantic cable until 1866, when the British *Great Eastern*, then the biggest ship in the world, laid a new and greatly improved cable. The first message was transmitted on September 7th, in the same year.

GRAMAPHONES AND RECORDS. *1877... the first gramaphone to play disc records. The turntable was rotated by hand. Invented by Emile Berliner.*

The record player

THE FIRST SOUND to be heard from Thomas Edison's phonograph was a squeaky little voice reciting *Mary had a Little Lamb*. That nursery rhyme was the

first sound ever to be recorded and preserved: all the speeches, musical interpretations and songs of the past before that time are lost to us.

Tin foil

Edison's invention seems remarkably simple. A needle was attached to a membrane of skin stretched across the narrow end of a horn. When Edison spoke into the horn his voice made the membrane vibrate. The vibrations then caused the needle to make tiny grooves and indentations on a sheet of tin foil which was wrapped round a cylinder turning by clockwork. The grooves were traced as a spiral round the tin foil.

"Frozen" sound

When this process was reversed and the needle was allowed to move along in the spiral groove, the indentations caused the needle to vibrate and thus the membrane also. In this way the sound 'frozen' in the grooves was reproduced and could be heard through the horn.

THE SECOND ATTEMPT. *Eight years after the first Transatlantic cable went dead, the crew of the "Great Eastern" tried again. On September 7, 1866, a new, and improved cable was successfully laid.*

That was in 1877. The first disc record and gramophone were invented ten years later by Emile Berliner, who greatly improved on Edison's phonograph by having the sound groove cut on a revolving disc instead of a cylinder.

In June, 1948, the Columbia Record Company called a press conference to announce the first long-playing record. The first British long-playing record was issued by the Decca Company in August, the following year.

The tape recorder

AT THE World Exhibition held in Paris in 1900 the major sensation was an instrument called by a strange name – the Telegraphon. It had been invented by a Danish engineer named Waldemar Poulsen and it was the true forerunner of the modern tape-recorder.

To capture sound on a gramophone record needed a lot of complicated and expensive equipment – not something the ordinary person could do at their leisure or in their own home. But the success of the phonograph and the gramophone encouraged inventors to experiment with a machine that would enable anyone, without technical knowledge, to reproduce any type of sound at any moment and to play it back immediately.

Poulsen's Telegraphon did just that: it took its recordings on a magnetizable wire which passed between the poles of an electromagnet and reproduced the original sound by reversing the mechanism. The serious disadvantage from which the Telegraphon suffered was that no recording made with it could be 'edited' to cut out unwanted parts of the recording. This was overcome in 1928, when Fritz Pfleumer patented a process which used instead of wire a magnetic tape.

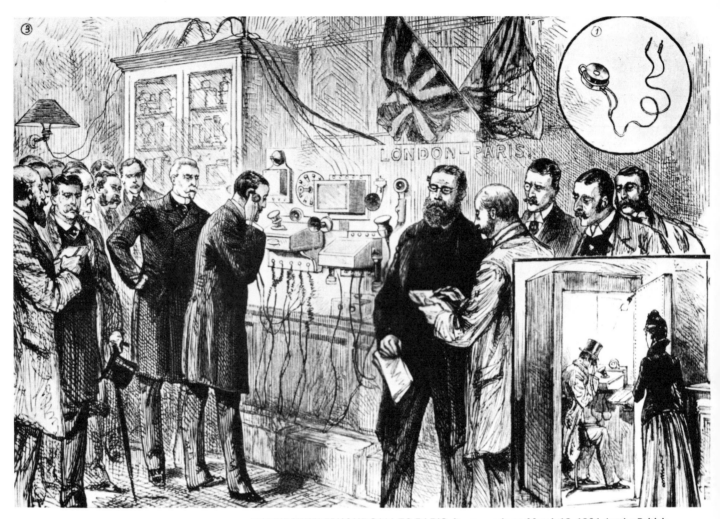

THE FIRST TELEPHONE CALL TO PARIS. *It was made on March 18, 1891, by the British Postmaster General, Mr Raikes, when he called M. Roches, the French Minister of Posts and Telecommunications.*

The telephone

THE FIRST telephone conversation was held on March 10th, 1876.

Alexander Graham Bell was a Scottish doctor who emigrated to America where much of his work was devoted to designing some kind of machine which would enable others to speak with people born deaf and dumb.

Bell based his experiments on the knowledge that when one speaks the air from the lungs makes the vocal chords vibrate. These vibrations are passed on to the molecules of gas that make up the air and cause vibrations in the air. If these vibrations – or sound waves – strike against a disc or diaphragm of thin metal, the disc will also vibrate back and forth.

In Bell's original telephone, the diaphraghms, were of steel, and vibrated near electromagnets, causing a small current. This process reversed itself at the other end; mouthpiece and receiver were identical.

It occurred to Bell that if he connected two diaphragms together by an electrified wire, words spoken against the first

A CALL FOR MR BELL. *When Alexander Graham Bell tried to design a machine that would enable people to speak with those who were born deaf and dumb, he ended up with this instead ... the first telephone.*

diaphragm would induce vibrations in the electric wire. These changes in the electric current passing through the wire would make the second diaphragm vibrate as it would if struck by sound waves.

The receiving diaphragm was to be held against the deaf-mute's ear. He or she would then have to be taught to read the meaning of the sound vibrations.

In his full-scale test, Bell mounted the mouth-piece of the apparatus on his laboratory bench and placed the receiver on a table in the room below. The two pieces of equipment were then joined by two wires.

In one room, Bell's assistant, a young man named Watson, picked up his instrument to listen. Suddenly, he heard Bell's voice, shouting, "Mr. Watson, come here. I want you".

Bell's shout was one of alarm as he had just spilled some acid on his clothes. He had also just made the first telephone call.

The electric telegraph

THE FIRST TELEGRAPH LINE was set up in 1838 along the London-Blackwall railway to control the movement of trains. One of the men responsible for developing this new method of communication was a soldier.

While on leave from his regiment in India, William Fothergill Cooke paid a visit to Heidelberg. That was in 1830. There he became friendly with Professor Muncke, who held the chair of physics at the university. During one of their meetings Professor Muncke demonstrated to Cooke that a magnetic needle could be made to swing backwards and forwards when it came under the influence of changes in an electric current.

Cooke resigns

It was six years later, when he was again on leave, this time in England, that Cooke was introduced to Charles Wheatstone, Professor of Physics at King's College, London. During their conversation Cooke discovered that Wheatstone had also been working on the idea of an electric telegraph.

Cooke promptly resigned from the army and entered into partnership with Wheatstone to construct an electric telegraph. Their joint experiments proved successful. The railway telegraph line was introduced.

Railway messages

Two years later they began installing their telegraph along the Great Western Railway line from Paddington. Like the London-Blackwall system, the Paddington telegraph was used solely for railway messages – not something to capture the imagination of the public. *That* happened dramatically on January 1st, 1845, when the receiving instrument at Paddington Station suddenly tapped out this message:

'*A murder has just been committed at Salthill and the suspected murderer was seen to take a first-class ticket for London on the train which left Slough at 7.42 p.m. He is wearing a brown greatcoat and is in the last compartment of the second first-class carriage.*'

When the 7.42 from Slough drew into Paddington Station police were waiting and the murderer, a man named Tawell, was easily arrested. He was eventually tried, found guilty and executed.

The incident aroused so much public interest in the possibilities of the electric telegraph that the following year Cooke and Wheatstone formed the Electric Telegraph Company for the transmission of messages by the public. Within a couple of years telegrams were recognised as an absolute necessity for the commercial and social welfare of the country.

LONDON'S FIRST TELEPHONE EXCHANGE. *It stood in Colman Street. It was the headquarters of the United Telephone Company. This picture of it appeared in "The Graphic" of September, 1883.*

Radio

RADIO WORKS by a series of electrical impulses, called electro-magnetic waves, which pass from one place to another without any wires connecting the transmitter and the receiver. No one actually *invented* radio: it is impossible to say at any one point, 'This was the *first* radio'. Its development was the result of the work of many scientists and inventors. But it is possible to trace steps which themselves were the first of their kind.

The first step in the development of radio was taken by the British scientist James Clerk Maxwell, who, in a lecture he gave to the Royal Society in 1864, suggested that when an electric current is passed through a wire or other conductor of electricity the current produces around it electrical disturbances which are thrown off in the form of invisible electro-magnetic waves.

Jumping sparks

Maxwell's theory was little more than an idea supported by a lot of complicated mathematics. He made no attempt to produce an apparatus for the specific purpose of creating the waves he talked about. But he did predict that some day the waves might be used for telegraphing messages without wires.

Next step

The next step in the development of wireless, or radio, was taken by the German scientist Heinrich Hertz. Twenty-four years after Maxwell's lecture, Hertz discovered that if an electric current was passed through an induction coil ending in two metal knobs, a spark would jump across the space between the knobs.

Hertz then fixed up a ring of copper some distance from his induction coil. The ring was not continuous but, like the coil, its two ends were topped by metal knobs which were separated by a tiny air space. When he switched a current through the coil, a spark once again jumped across the air gap between the coil's two knobs; at the same time an exactly similar spark jumped across the air gap in the metal ring.

Speed of light

As there was no wire or other connection between the coil and the ring, Hertz decided that the spark on the coil had set up electro-magnetic waves which had carried the spark to the ring. He even worked out the speed of the waves – 186,000 miles a second: the speed of light.

Variable oscillations

Then he carried his experiment a stage further. He arranged a device that could change the oscillations, or number of cycles, of the current that produced the spark. The spark on the ring followed suit and changed accordingly. It seemed to be an exciting discovery, but curiously enough Hertz was not particularly interested beyond the initial experiment. He made no attempt to exploit his discovery.

No attempt either was made to exploit the concept by Sir Oliver Lodge, who, in 1894, made a convincing demonstration of electro-magnetic waves at a meeting of the British Association.

On Sir Oliver's desk in the lecture hall was placed a Hertz induction coil together with a device for changing the oscillation of the electric current which was to be passed through it. The whole apparatus was now called an 'oscillator'. In another room, 100 feet away and separated by two stone walls, was a Hertz ring or, as it had come to be called, a 'resonator'.

THE KITE AERIAL AT SIGNAL HILL. *In England, Marconi soon found distrust and disinterest in official quarters. The Post Office withdrew support but a group of business men backed him to develop radio on a large scale. On December 12, 1901, Marconi sent out three Morse dots – which were picked up on Signal Hill, Newfoundland, the world's first Transatlantic radio transmission.* ABOVE, *Marconi, left, supervises the erection of the kite aerial at Signal Hill.*

RIGHT: THE FIRST ADVERTISED BROADCAST *of entertainment in Britain. In 1920, two years before the B.B.C. was officially inaugurated, Dame Nellie Melba broadcast a recital of songs.*

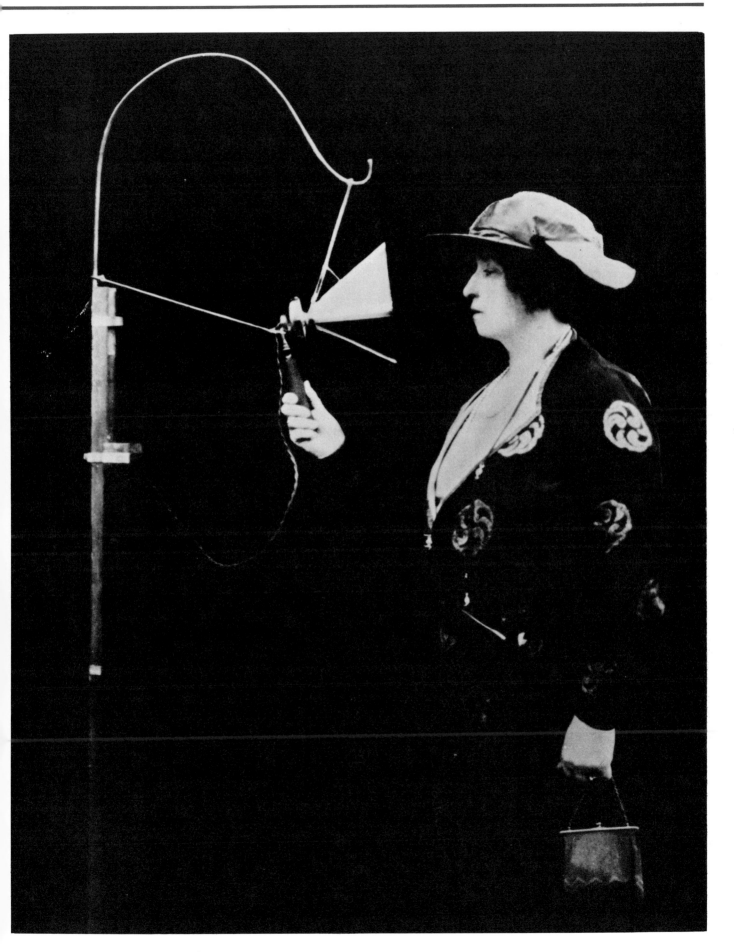

The leaping spark

When Lodge switched on his oscillator, the spark that leapt across the gap in the coil was repeated, as before, across the gap on the resonators – even at that distance.

Although Sir Oliver did suggest that by varying the current in the oscillator it might be possible to send messages in Morse code with the apparatus (he had already shown that the needle of a galvanometer – for detecting the presence of an electric current – attached to the resonator could be moved by changing the current from the oscillator), nonetheless he did not make the experiment himself.

Poppof takes over

The first Morse signal to be sent by radio was achieved by a Russian scientist, Alex Popoff, in 1895, using equipment very similar to that used in the experiments by Hertz and Lodge. The distance between Popoff's transmitter and receiver was three miles. Once again, no

THE FIRST MARCONI TRANSMITTER.
This is the transmitter Marconi brought from Italy to England in 1896. The British Post Office backed him at first, and he set up a transmitter on Salisbury Plain.

attempt was made to exploit the idea commercially.

This was left to an Italian – Guglielmo Marconi, the most famous name in radio. Marconi had followed the earlier experiments and improved on the apparatus that had been used. In the same year that Poppof sent his code signal, Marconi succeeded in making an electric bell ring by radio transmission, and a few weeks later transmitted a message in Morse across a distance of *five* miles.

Businessman

However, unlike the earlier pioneers, Marconi was not only a clever electronics engineer, he was also a keen business man and tried immediately to interest the Italian government in the possibilities of radio as a means of commercial communication. When the government refused to provide any money, Marconi went to England, where, with the help of the Post Office, he set up a transmitter on Salisbury Plain and in March, 1897, succeeded in sending a short message in Morse to a receiver on the roof of the General Post Office headquarters in London.

Too much money

The Post Office then decided that Marconi was spending too much public money, and withdrew its support, which was replaced by a group of business men who saw the potential of radio: on July 27th, 1897, the Marconi Wireless Telegraphy Company was formed with the Italian as technical director. At once he set about developing radio on a large scale.

For a start radio was installed on a number of ships and a receiver and transmitter were set up on the South Foreland Lighthouse, in Kent. It was from there that the first S O S, or distress call, was sent out, in 1899, when the lighthouse keeper saw two ships collide and the radio was used to call the lifeboat.

The greatest triumph

On December 12th, 1901, came Marconi's greatest triumph: a transmitter at Poldhu, Cornwall, signalled the three dots in Morse code representing the letter 'S'; the signal was picked up by a receiver on Signal Hill, Newfoundland – the first Transatlantic radio transmission.

Three years later, Marconi started the first radio news service between Britain and the U.S.A., and three years later still the first commercial radio telegraph ser-

vice across the Atlantic was begun. The year before that, in 1908, the Post Office opened the first ship-to-shore radio station at Holt Head, in Devon.

● In December, 1906, R.A. Fessenden made the first experimental broadcast of violin music and singing from Brant Rock in Massachusetts. The broadcast was picked up at a distance of 211 miles. He had made a short broadcast of speech in November, six years earlier, but it had been too distorted to be of value.

● The first transistor was produced in 1948 by William Shockley, Walter Brattain and John Bardeen, all scientists at the Bell Telephone laboratories in America.

Twenty-two years to invent the microphone

MARCONI TRIES TO INTEREST ITALY. *By the mid-1850's, the Italian, Guglielmo Marconi, had pushed the development of radio onwards. He succeeded in making an electric bell ring by radio transmission and in sending a Morse signal a distance of five miles. He demonstrated his work to members of the Italian Government (picture above) but they refused to back him.*

The microphone

DAVID HUGHES invented the microphone in 1878. He had been experimenting with microphones for 22 years. His early ideas were unsuccessful.

The simple solution on which he eventually hit was to fill a small cylinder with loosely-packed carbon granules – carbon is one of the non-metallic substances that best conducts electricity.

At one end of the carbon-filled cylinder there was a thin piece of metal called a diaphragm; a wire passed from the diaphragm through the carbon granules to a further 'hearing' diaphragm. An electric current was passed through the wire. The sound waves vibrated against the first diaphragm; the vibrations affected the carbon granules, which in turn affected the vibrations against the 'hearing' diaphragm.

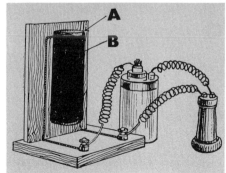

FOR 22 YEARS, *David Hughes met failure after failure in his attempts to design the first microphone, shown above. Finally, in 1878, he hit on the simple solution.*

THE WORLD'S FIRST RADIO TELESCOPE, *was built at Jodrell Bank, England. Its super-sensitive receiver picks up signals from stars either too far away, or not bright enough, to be seen.*

Unseen stars

All matter, including the stars, is made up of particles which are constantly moving. In doing so the particles release radiation in much the same way as a broadcasting station transmits its programmes. Compared to a broadcasting station, however, the strength of the signals caused by the radiation in the stars is very low, but given the right kind of receiver, tuned correctly, the signals can be picked up.

The telescope at Jodrell Bank was specially designed to do just that. It

originate in the Milky Way.

During the next fifteen years ghost signals were picked up by stations all over the world. Scientists did not for a moment believe that the signals were being transmitted from intelligent beings: they decided that the transmissions must be from what are called radio stars.

The radio telescope

THE WORLD'S FIRST radio telescope went into operation at Jodrell Bank, Cheshire, in 1957.

Twenty-five years earlier electrical engineers tracing a fault in the Transatlantic radio-telephone service had picked up some weak and weird signals in their receivers, signals that were not in any known code.

Careful checking proved that the mystery signals were not transmitted by any station on earth, but seemed to

themselves run on bogies on a circular track so that the aerial can be pointed in any direction. Signals from the telescope are turned into electrical impulses that in turn drive a pen that draws a graph of the broadcast. From these graphs astronomers have gained a mass of new information about conditions in outer space.

Radio telescopes have picked up signals which must have been travelling through space for thousands of millions of years. If that is the case, then the stars from which the signals set out may no longer exist.

Space communication

THE FIRST COMMUNICATION between earth and space took place on October 4th, 1957, when the Russians opened the gateway to space travel by launching Sputnik I, the first man-made satellite to enter orbit round the earth.

The unmanned metal sphere weighed 184.3 pounds and was 23 inches in diameter. It orbited the earth in 96 minutes and made approximately 1400 orbits in all. On re-entering the lower atmosphere it burnt up.

Sputnik's purpose was to measure and transmit information on the density and temperature of the upper atmosphere as well as to measure the concentration of the electrons in the ionosphere. Signals from its radio transmitter were picked up all over the world.

A communications satellite

TELSTAR WAS THE first satellite to be used for bouncing a television picture around the world.

Television broadcasts have to be transmitted on very short waves, in the U.H.F. waveband. Unlike the long waves used for speech and morse broadcasts, U.H.F. waves are not reflected back to earth by the layer in the atmosphere known as the ionosphere. Therefore U.H.F. cannot be bounced from one side of the world to the other: instead, they travel in an almost straight line across the curvature of the earth and into space. Normally the greatest distance at which a television broadcast can be received is the horizon within sight of the transmitting station.

18,000 m.p.h.

Telstar provided the necessary object in space from which the U.H.F. could be bounced to greater distances around the world.

The satellite, weighing 170 pounds, was launched from Cape Canaveral (now Cape Kennedy) on July 10th, 1962, and put into orbit round the earth at a maximum height of 3,000 miles and a minimum of 600 miles, to travel at a maximum speed of 18,000 miles an hour.

Lunar communication

MAN'S FIRST CONVERSATION between the moon and the earth took place in 1969, when the American astronauts Neil Armstrong and 'Buzz' Aldrin landed on the moon. They broadcast to stations on earth speech and television accounts of what they saw and were doing. Throughout the whole of their stay on the moon the astronauts were in constant two-way communication with earth.

consists of a super-sensitive receiver which can pick up signals from stars either too far away or not bright enough to be seen on an enormous aerial shaped like the reflector bowl of an electric fire.

New information

The aerial is mounted on towers and can be moved up or down; the towers

RIGHT: 1957 . . . MAN'S FIRST SPACE SHOT. *Sputnik 1, twenty-three inches in diameter, 184 lbs in weight. This was Man's first satellite. It made 1,400 orbits of the earth.*

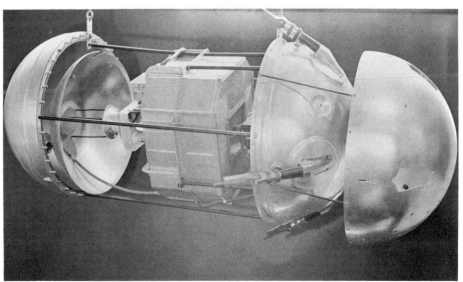

Part 8

SPORT

*Man has been able to dream up a great many different ways of
exercising himself and enjoying his leisure. We have
included a few here of these sports to show their variety.
They show, too, some of the extraordinary number of variations
of games that can be played with a ball or with a bat and ball
of one shape or another – many of the games so old that we
cannot trace their origins.
One might have thought that every possible variation had been
tried out long ago but there has been at least one new ball game
that was invented within the last hundred years: modern basketball.
And yet it seems such an obvious game that one would have been
prepared to bet that it had been in existence for thousands of years.
So perhaps within the next hundred years someone else will
think up yet another ball game that is completely new.*

SOUTHWARK PRISON, LONDON. *The inmates enjoy a game
of rackets. By 1820, this prison-born game had spread across
continents. Later, the new game of squash developed from rackets.*

Football

NOBODY INVENTED FOOTBALL. The Egyptians, Assyrians and Greeks all kicked, bounced or rolled objects around and, from this simple base, emerged Association, Rugby Union, Rugby League, American, Gaelic, Australian Rules and various other kinds of football.

The game is thought to have come to Britain with the Romans, and has been traced back to AD 217 in Ashbourne, Derbyshire. The first formally constituted club was the Sheffield Club in 1857, six years before the Football Association itself was formed. Soon afterwards the clubs that adhered to Rugby School rules (i.e. where handling and hacking were allowed) broke away and later formed the Rugby Football Union.

The Football Association first organised its famous Cup competition in 1871, and the first winners, in 1872, were a club of Harrow School old boys, the Wanderers.

Preston Undefeated

By 1888 the Football League had come into existence, its first winners being Preston North End who did not lose a single game during the 1888-89 season

THE FIRST F.A. CUP. *The Football Association Cup competition was started in 1871. This was the first F.A. Cup. It was stolen in 1895 and has never been found.*

and remain the only club in the English game's history to have gone through a season undefeated and the first to win the double of both League and Cup.

The first World Cup was not played until 1930, the winners being Uruguay. None of the four British countries entered before 1950, and England first won the trophy in 1966. The European Cup was first won by Real Madrid in 1956 and held by them until 1960. The first British winners were Glasgow Celtic in 1967.

Rugby

RUGBY IS ONE of the many variations to emerge from the basic game of football and it is, perhaps, the closest to its progenitor. The original aim of a mediaeval football match was simply to carry the ball to your opponent's goal – this usually being one end of a town or the end of a field. Modern rugby is surprisingly similar and owes its form to one William Webb Ellis. He was a pupil at Rugby School and in 1823 simply picked up the ball and ran with it to his opponents' goal-line. At that time footballers were allowed to handle the ball when stationary but not run while holding it. The clubs that preferred rugby broke away from the Football Association and formed the Rugby Football Union in 1871.

The first rugby international was arranged between England and Scotland the same year. As the game has remained strictly amateur there were no organised club competitions until 1972, when Gloucester won the first official cup competition sanctioned by the RFU.

The break-away

In 1893, however, a group of northern clubs suggested that payment should be made to players in the tradition of Association Football. This was rejected by the south, and the northern clubs finally broke away to form the Northern Union (later commonly known as the Rugby Football League) on 29 August 1895. Their rules gradually diverged from those of Rugby Union, including the use of 13 instead of 15 players, and nowadays the games have many differences.

The major club competitions are the Rugby League Challenge Cup, the first winners of which were Batley in 1897, and the Northern Rugby Football League, first won by the now long-defunct Manchester club Broughton Rangers in the 1901-02 season.

There is also a Rugby League World Cup held every few years between the four major countries – Australia, New Zealand, France and Britain. Britain were the first winners in 1954.

Rackets and squash rackets

RACKETS IS AN EXTREMELY fast game for two or four people, played in a plain four-walled court with rackets and a small, hard, ball. The origin of the game is uncertain, though it is believed to have

The game the Romans brought to Britain

begun in Fleet prison, London.

In 1820, Robert Mackay, of London, was recognised as World Champion in that he 'claimed' the title. Forty years later, Francis Erwood became the first Closed Court Champion.

Squash Rackets developed from Rackets. In 1850, a group of boys who were waiting to play rackets in the court at Harrow School began knocking up against the outside walls of the court – and started a habit.

Their Housemaster quickly banned the use of the hard ball outside the court.

Not to be outdone, the boys invented a game of rackets played instead with a soft, or 'squashy', ball. Various other small changes were made, and the game of Squash Rackets was born.

Court sizes

A Squash Rackets Association was founded in 1907 in the United States using a court of different dimensions from the British one.

The first Professional Championship in Britain was held in 1920 and the first Amateur Championship was held in

THE GREEKS DID IT. *The Assyrians, the Egyptians and the Romans did it. And, many authorities believe that it was the Romans who brought the game of football to Britain. Certainly it can be traced back clearly to Ashbourne, Derbyshire in A.D. 217. This picture shows football as it was in London about 1307.*

1922. Home internationals began in 1937.

The game is immensely popular in Egypt, Pakistan and India, and has a growing following in South Africa, Australia and, in recent years, in Japan.

ABOVE: THE MAN WHO BROUGHT LAWN TENNIS TO ENGLAND. *Major Walter Wingfield introduced "lawn tennis" at a garden party in 1873. He wanted to call it "Sphairistike", which is Greek for "Play". This was too difficult for the English devotees who preferred to call it "tennis-on-the-lawn".*

LEFT: THE GAME A KING BANNED. *The Royal and Ancient Golf Club of St. Andrews, Scotland, the home of golf. This is it in 1880. But, as far back as 1440, King James II of Scotland tried to ban golf by law because his subjects were giving up the compulsory sport of archery to follow the little white ball. His efforts were completely unsuccessful. Later, Queen Mary of Scotland herself founded St. Andrews, the first royal golf course.*

HE TRICYCLE SKATE. *A Mr J. Walters of London dreamed up the icycle skate in the 1880's. Twenty miles an hour, said Mr Walters. idiculous, impossible, said the "Queen" magazine, but a aluable addition to our means of "locomotion". Seventeen years arlier, in fact, James Plympton of New York had invented the ur-wheeled roller-skate. And this is the type that has survived, ven if it is rather less entertaining to look at than the tricycle kate!*

TOP: THE FIRST BASKETBALL TEAM *was formed of students of what is now Springfield College, Massachusetts. Dr Naismith, the game's inventor, was an instructor there.*

MIDDLE: DIOMED *winner of the first Derby, held in 1780. The race is for three-year-olds and is run over one and a half miles.*

BOTTOM: IN 1892, *an American patented the first "ski-bike". This is a ski-bobber (as ski-bike riders are called) of 1947. The first known ski-bike was patented as early as 1892, however.*

1690 . . . when ladies played the first billiards

THE GAME THAT BEGAN 2,500 YEARS AGO. *The Persians played polo in 525 B.C. The aristocracy of India played it. British cavalry officers there adopted it and brought it to England in 1869. In 1874 the famous Hurlingham Club was formed – this is the scene there, in 1915.*

Billiards

BILLIARDS WAS A French invention that very quickly caught on among the Germans, the Dutch and the Italians. But we cannot be sure when the French first played the game. At any rate, as it is mentioned by Shakespeare, billiards must have been known to the English in the sixteenth century.

The first public billiards room in England was the Piazza, Covent Garden, which was opened in the early nineteenth century. The rubber cushion was introduced in 1835 and slate was used to give the table a true surface in the following year.

In France the tables are made without pockets and the game consists entirely of cannons. In England the table has six pockets in order to give the game greater variety. Other billiard-table games are snooker and pool.

The balls themselves were first made of the finest ivory. Composition balls are used now.

Polo

POLO IS QUITE POSSIBLY one of the oldest organised sports. It was first played in the East and the name is believed to come from a Tibetan word, 'Pulu', meaning 'ball'.

The first known mention of the game was by King Darius of Persia as long ago as 525 B.C. Its popularity spread along the trade routes of the ancient world and it was played by the aristocracy of India. It was in India, during the time of the British Raj, that British cavalry officers learnt and adopted the game.

The ponies are bred specially for the game. Fresh ponies are brought in during the game and much depends on the animal's swiftness and agility, but still more on the skill with which the riders manage them.

The first club

The first polo club was founded in Assam, in 1859. Ten years later, the game was introduced into England by the Tenth Hussars, at Aldershot. The first polo club in England was formed in 1872. A set of rules was drawn up at the same time.

The famous Hurlingham Club was formed two years later, and became the headquarters of polo in England. The first international match between Britain and the United States, for the Westchester Cup, was played in 1886.

Yachting

THE EARLY EGYPTIANS and the Romans used pleasure craft which one might call 'yachts', but yachting as a sport in the modern sense seems to have originated in Holland. The word itself comes from the Dutch word 'jaght'.

Yachting was introduced into England by Charles II on his return from exile in Holland. He raced his brother James, Duke of York, for £100 a side on September 1st, 1661, over a 23-mile course from Greenwich to Gravesend.

The earliest known club in Great Britain is the Royal Cork Yacht Club (formerly the Cork Harbour Water Club), which was established in 1720. The Royal Thames Yacht Club followed as the first in England.

LEFT: HOW BILLIARDS WAS BORN. *The French were the first to play it. Here, in 1690, a lady plays a tricky shot.*

RIGHT: THE FIRST YACHTS *were probably Egyptian. But yachting as we know seems to have began in Holland.*

103

Boxing

FIGHTING WITH BARE KNUCKLES had been in existence for years before James Figg became the first recognised Champion of England in 1719. He held the title for about 11 years.

James Figg claimed to be Champion of the World, as well, but he never fought abroad or, so far as we know, in any kind of international competition. Nonetheless, he was the chief authority on boxing in his day and everyone followed his basic rule that fighters should continue until there was a definite winner – no rest periods allowed.

In 1743 Jack Broughton made radical changes to check the brutality of the sport. He drew up the first rules. The most important of these was that a round should continue until one or other of the fighters was knocked down, when he would be given 38 seconds to get back on his feet or be declared the loser.

No throwing

The rules which stand today and which were introduced by the Marquess of Queensbury did not appear until 1867 – 122 years later. It was then that gloves replaced bare fists. But it was not for another five years that all Queensbury's rules were adopted, including three minute rounds and the barring of throwing and wrestling.

In that year, 1872, fighters were classified into different weight categories for the first time: light-weight 133 pounds or below; middle-weight 154 pounds or less; heavy-weight above 154 pounds. In the same year, trophies were introduced for the first time: previously all fights had been for cash prizes or to settle a grudge.

Ski-ing

IT IS IMPOSSIBLE to tell when men first began to use skis to cross the snow. The earliest known ski was found in a peat bog in Sweden and is now in a museum at Stockholm. This ski is believed to be 4,500 years old. Skis must have been in use long before then.

The earliest picture of men on skis is a rock carving that was found in Norway: two hunters are ski-ing after elk. The carving was probably done by a Stone Age artist about 4,000 years ago.

The first skis were made from bones of large animals and strapped to the feet with leather thongs. A wooden pole was used for braking.

Skis were first used in warfare during the Battle of Oslo in A.D. 1200. Over 300 years later, in another war, the Swedes designed the first known stretchers by stretching animal skins between two skis and using them to transport their wounded.

Ski-ing, as we know it today, began with the invention of ski bindings in the 1880s by Sondre Nordheim, of Norway, though the first organised ski club was the Trysil Club in Norway, formed in 1861. Mathias Zdarsky published the first methodical analysis of how to turn on skis and, in 1896, developed the first ski specially designed to assist turning.

An Englishman, E. C. Richardson, initiated the ski-ing proficiency tests. He also helped to found the Ski Club of Great Britain, in 1903 – the world's first national administrative body for ski-ing.

The first Winter Olympic Games were held at Chamonix, in France, in 1924.

ABOVE: THE FIRST SKI OLYMPICS. *Ski-ing blossomed as an international sport in 1924, when the first ski Olympics were held at Chamonix.*

LEFT: *Major E. C. Richardson helped to found the Ski Club of Great Britain – and he ran the first proficiency tests.*

RIGHT: THE RULES AGREED "BY SEVERAL GENTLEMEN". *Broughton's boxing rules included a white chalked square in the middle of the ring – and a "rest period".*

RULES

TO BE OBSERVED IN ALL BATTLES ON THE STAGE

I. THAT a square of a Yard be chalked in the middle of the Stage; and on every fresh set-to after a fall, or being parted from the rails, each Second is to bring his Man to the side of the square, and place him opposite to the other, and till they are fairly set-to at the Lines, it shall not be lawful for one to strike at the other.

II. That, in order to prevent any Disputes, the time a Man lies after a fall, if the Second does not bring his Man to the side of the square, within the space of half a minute, he shall be deemed a beaten Man.

III. That in every main Battle, no person whatever shall be upon the Stage, except the Principals and their Seconds; the same rule to be observed in bye-battles, except that in the latter, Mr. Broughton is allowed to be upon the Stage to keep decorum, and to assist Gentlemen in getting to their places, provided always he does not interfere in the Battle; and whoever pretends to infringe these Rules to be turned immediately out of the house. Every body is to quit the Stage as soon as the Champions are stripped, before the set-to.

IV. That no Champion be deemed beaten, unless he fails coming up to the line in the limited time, or that his own Second declares him beaten. No Second is to be allowed to ask his man's Adversary any questions, or advise him to give out.

V. That in bye-battles, the winning man to have two-thirds of the Money given, which shall be publicly divided upon the Stage, notwithstanding any private agreements to the contrary.

VI. That to prevent Disputes, in every main Battle the Principals shall, on coming on the Stage, choose from among the gentlemen present two Umpires, who shall absolutely decide all Disputes that may arise about the Battle; and if the two Umpires cannot agree, the said Umpires to choose a third, who is to determine it.

VII. That no person is to hit his Adversary when he is down, or seize him by the ham, the breeches, or any part below the waist: a man on his knees to be reckoned down.

As agreed by several Gentlemen at Broughton's Amphitheatre,

Tottenham Court Road, August 16, 1743.

Tennis

TENNIS ORIGINATED AS a ball game in fifteenth-century France. At first it was played by hand, using a net *and* a wall; then a *battoir* was used – a bat shaped rather like a paddle. The name comes from the French command, '*Tenez!*' ('Hold!'), because the racket had to be held firmly when striking the ball.

The French kings Louis XI, Henry II and Charles IX were expert and enthusiastic players and because of its popularity in the French courts the early form of the game was known as 'Real (*Royal*) Tennis'. Henry VIII of England also played Real Tennis and even had a court built at Hampton Court.

The first book on Real Tennis was written by an Italian, Antonio Scano de Salo, and published in 1555; the first book of rules was published by a Frenchman, Forbet, in 1592.

The Major's role

Lawn Tennis was derived from Real Tennis. It was introduced by a British Army officer, Major Walter Wingfield, at a garden party in 1873. He claimed that the game had been known to the Ancient Greeks, and patented the name '*Sphairistike*', which is Greek for 'Play'.

This was too difficult a name for English players, who preferred to call it 'tennis-on-the-lawn'.

The first English championships were held at Wimbledon in 1877 and were won by Spencer Gore. In 1900 an American, Dwight F. Davis, put up a cup for a contest between Britain and the U.S.A. It was played at Boston, Massachusetts, and the Americans won 3-0. Later, the competition was opened to all nations.

Golf

THE GAME IS SO OLD that the origin of it is argued about not only among sportsmen but even among archaeologists.

Some insist that it was first introduced in Scandinavia or Northern Germany but, until 100 years ago, there is no satisfactory record of golf being played anywhere outside Scotland, where it has been popular since 1440. Indeed, it was so popular then that King James II of Scotland passed a law forbidding his subjects to play, because they were becoming more interested in golf than in the compulsory sport of archery.

Clearly the law did not hold force for long. Mary, daughter of James V, learned the game in early childhood and later referred to players as 'cadets', meaning pupils, which is where we get the present term 'caddy' from. When Mary became Queen of Scotland, she founded St Andrews – the first royal course and still the headquarters of the game. In 1834, William IV became patron of St Andrews and conferred on it the title of 'Royal and Ancient'.

First champion

The first recorded tournament was played in 1860 at the Prestwick course, where the British Open Championship has often been held in recent times. The first recognised Open Champion was Willie Park.

The rubber-filled ball was invented in America in 1899 and introduced in 1902.

Roller-skating

THE FIRST ROLLER-SKATE was designed by Joseph Merlin of Huy, in Belgium, in 1760 – more recently, perhaps, than one might have imagined, though Merlin's roller-skate was probably not very efficient.

Improved versions appeared during the next 100 years, but a satisfactory roller-skate was not designed until 1863, when James Plympton of New York made the four-wheeled type, patented it and opened the first public rink in the world

ROLLER-SKATING ON THE ROADS.
This Dutch skater was well-known around the streets of the Hague, in 1790.

at Newport, Rhode Island, in 1866.

Basketball

THERE IS NO DISPUTE about the origin o modern basketball: two peach basket fastened to a gymnasium balcony in th winter of 1891 launched a new an

THE WIMBLEDON CHAMPIONSHIPS. *The first English championships were held at Wimbledon in 1877. This picture shows the fifth round in the fifth year of the tournament. In 1900, an American, Dwight F. Davis, put up a cup for a U.S.A. v. Britain contest. Later the contest was opened to all nations.*

rapidly popular sport.

The inventor of the game was Dr James Naismith, an instructor at the International Young Men's Christian Association school at Springfield, Massa-chusetts – now Springfield College. It is remarkable that 12 of the 13 rules originally framed by James Naismith still exist in the game and are basic to it, because the game itself has changed to some extent. The peach baskets soon gave way to metal ones, then, in 1906, to open hoops.

Professional basketball was launched on an organised basis in 1892. The National Basketball League, as it was known, had teams in New York, Philadelphia, Brooklyn and Southern New Jersey.

Motor-racing

THE FIRST 'AUTOMOBILE TRIAL' covered 20 miles from Paris to Versailles and back, on April 20th, 1887. The trial was won by Georges Bouton's steam quadricycle in 74 minutes – an average speed of just over 16 miles an hour.

The first full-scale race was from Paris to Bordeaux and back – a distance of 732 miles. It was held between June 11th and 13th, 1895. The winner was Emile Levassor, from France, driving a Panhard Levassor two-seater with a 1.2 litre Daimler engine which developed up to four horsepower. Emile Levassor's time was 48 hours and 47 minutes – an average of almost exactly 15 miles an hour.

Cricket

SERIOUS HISTORIANS OF CRICKET insist that the game originated and was developed in England but they do not know the date when it was first played, where it was first played, or how it gained its name.

There are some historians who have shocked the cricketing world by suggesting that the game is derived from croquet, which has always been popular in France. These same historians have even dared to suggest that the English borrowed the idea from the French, altered it to suit their own national needs and finally perfected it into modern cricket.

Others prefer to believe that cricket came from 'club ball', an early English game so old that details of how it was played are lost. Anyway, a drawing in the King's Library, London, establishes that cricket was a well-organised sport as early as 1344.

An enquiry

A record of the Quarter Sessions in the archives of Guildford Corporation contains what is thought to be one of the earliest written references to the playing of cricket. This record includes the minutes of an enquiry held in the year 1597–98 concerning the wrongful enclosure of a piece of land. The record states:

'One of the Queen's (Queen Elizabeth I) *coroners of the county of Surrey gave evidence that, when he was a scholler in the free schule of Guildford, he and several of his fellows did run to play there at crickett . . .'*

In the earliest versions of the game, there were no wickets: the players cut circular holes in the turf. A batsman was 'out' when the ball was thrown in to the hole before the bat was grounded in the crease – a line drawn four feet in front of the hole. A type of wicket was added about 100 years after the reference to the game at Guildford.

The first recorded county cricket match was played between Kent and Sussex in 1728. Sixteen years later, the London Cricket Club sponsored a revision of the rules that governed the game. In 1777, three stumps were introduced for the first time: they measured 22 inches by six inches. Three years earlier, the weight of the cricket ball had been fixed at $5\frac{1}{2}$–$5\frac{3}{4}$ ounces, the width of the bat at $4\frac{1}{4}$ inches.

Modern cricket certainly dates from an historic meeting in London of the Marylebone Cricket Club in 1788, when all previous rule-books were revised and the present basis of the game was firmly

established. The MCC was first known as the Artillery Ground Club; later this was changed to the White Conduit Cricket Club; it was changed again – to its present name – in 1787.

Scoreboards were used for the first time at Lord's Cricket Ground in 1846. Thirty years later, the first England v Australia Test series was played in Australia.

The Ashes, the legendary cricket trophy held by the winners of a Test series between England and Australia originated in a mock obituary notice that appeared in the *Sporting Times* in 1882. England had just lost the only Test of the

WHO STARTED CRICKET? *This question causes endless discussion but it is certain that the first recorded cricket match was played between Kent and Surrey in 1728. The above picture shows a game in 1770. In 1777, three stumps were introduced for the first time.*

The first car race . . . zooming along at 15 mph

season by seven runs, after being set only 85 to win. The notice read as follows:

In affectionate Remembrance
of
ENGLISH CRICKET
which died at the Oval
on
29th August, 1882
Deeply lamented by a large
circle of Sorrowing Friends
and Acquaintances
R.I.P.
N.B. The body will be cremated
and the Ashes taken
to Australia.

THE WINNER. *The first full-scale motor race was from Paris to Bordeaux. Here is the winner, Emile Levassor, racing home at 15 m.p.h.*

When England, captained by the Hon. Ivo Bligh, later Lord Darnley, won in Australia in the following (southern) summer, some ladies burnt a stump, put the ashes in an urn, and presented it to the England captain. Lord Darnley bequeathed it to the MCC in his will. The urn does not change hands, but remains in the Memorial Gallery at Lord's. So the Ashes for which the two countries play are really an imaginary trophy.

Ski-bobbing

SKI-BOBBING WAS a winter pastime that became an organised sport in the 1950s. A ski-bob looks like a bicycle with short skis where there should be wheels. The rider also wears miniature foot-skis fitted with metal claws at the heel to assist braking. Experts have been known to reach speeds of 100 miles an hour on ski bobs.

The first known 'ski bike' was patented by an American called Stevens, in 1892. The first official world championships for ski-bobbing were held in 1967, at Bad Hofgastein, Austria.

Baseball

AMERICANS THEMSELVES DISAGREE with each other about the origin of their national sport.

Some attribute it to Abner Doubleday, who became a general in the U.S. Army. He is supposed to have laid out the first baseball diamond at Cooperstown, New York, in 1839, and to have formulated a set of rules. Others argue that baseball was merely an adaptation of the old English games of rounders and cricket.

In any case, it has been known by many names: 'Town Ball' in Philadelphia and 'One Old Cat' in Boston. In its early days, scoring terms and references were very similar to those used in cricket. So perhaps there is some likelihood in that origin.

1858 . . . the professionals

We do know that in 1841 New Yorkers decided that the rules should be re-organised. Rules were first modified in 1845 by Alexander Cartwright of New York. An early match on record was the New York Nine's defeat of Cartwright's

New York Knickerbockers by 23–1 on June 19, 1846, in Hoboken, New Jersey.

Until July 20, 1858, all baseball players were amateurs and there was no admission fee for the games. On that date, the first professionally promoted game was held on the Fashion race track, Long Island, between New York and Brooklyn stars. The gate raised 750 dollars.

There was no standard ball until April 27th, 1938, when a regulation National League ball was introduced.

Horse-racing

NO DOUBT PEOPLE HAVE BEEN racing on horseback for thousands of years. Roman chariot-racing brought some order to the sport and perhaps laid the basis for our modern form of horse-racing. Four miles was then the standard distance for the Olympiads.

The first public racecourse in Britain was the Smithfield Track, which was built about 1174, in London. This is regarded as the birth-date of organised horse-racing under saddle. Mostly, then, men raced for glory, or some private prize perhaps.

In 1512 the promoters of Chester Fair offered the first official horse-racing trophy of which we know – a wooden ball bedecked with flowers as a prize to the winner. The tradition was carried on for a long time.

Two hundred years later, Queen Anne introduced the sweepstakes – racing for a cash prize. She herself was an enthusiastic race-goer and encouraged horse-racing by presenting a gold cup for the winner of the main race at Doncaster. In 1714 she changed the rules by declaring that the owners of all the starters in a race should each put up 10 guineas – winner take all. It may not have been surprising that her own horse, Star, no doubt a hot favourite, won the first such race in history.

In 1780 the 12th Earl of Derby established a race to be run over one-and-a-half miles, for three-year-old horses. He did not give the race a name, but the public quickly christened it Epsom's Derby and, under that name, it has become the greatest horse-race in the world. The first winner was Diomed, owned by Sir Charles Bunbury.

Steeplechasing

IN THE EIGHTEENTH CENTURY there were plenty of contests such as the earliest recorded cross-country race, held in Ireland in 1752 between O'Callaghan and Blake, who matched their horses to race four-and-a-half miles between Buttevant Church and St. Leger Church.

There was no fixed course and the cry was 'Devil take the Hindmost', but the steeple of the church was the most prominent landmark and an obvious finishing post: 'steeplechasing' took its name from that.

Forty years later, in Leicestershire – a great hunting county – there was a

THE GREATEST RACE IS BORN. *This is the first Derby, in 1780, it was established by the Earl of Derby as a race to be run over one-and-a-half miles, for three-year-old horses. The public called it "Epsom's Derby".*

match for 1,000 guineas between two professional horsemen, Willoughby and Hardy. That was possibly the first race in which professional 'jockeys', in the modern sense, took part.

Gentlemen only

In 1811 the first race over fences was held at Bedford. Fugitive beat Cecilia over a three-mile course in which four four-foot-six-inch fences were each jumped twice. Nineteen years later the St. Albans Races were launched: before that date most races were matches between gentlemen riders only.

In 1836 the first races were run at Liverpool, including a 'Sweepstakes of 10 sovereigns each with 100 sovereigns added by the town of Liverpool'. Some authorities regard the corresponding race as the first Grand National. There were only four runners but the favourite had come over from Ireland, so we may

suppose that it was considered an important event. Certainly it helped to make Liverpool the early centre for the sport. The winner on that first occasion was The Duke.

THE MAN WHO CREATED "THE DERBY". *The 12th Earl of Derby did not know that the race he established would become the greatest horse race in the world.*

Ice-skating

THE FIRST KNOWN ICE-SKATING club was formed in Edinburgh in 1742, although Samuel Pepys – famous for his diary – had earlier described skating on the frozen lake in St. James's Park, London, in 1662, and had taken advantage of the great frost of 1683 to dance on the ice with Nell Gwynn.

The technique of ice-skating is believed to have originated in Scandinavia 3,000 years ago, when it was used as a means of transport. It is certainly mentioned in Scandinavian literature of the second century A.D.

In the Netherlands, there were skaters on the canals in the Middle Ages: a Dutch wood engraving of 1498 shows St. Lydwina of Schiedam, who broke a rib while skating in 1396. She died 37 years later – so the accident could not have been that serious – and became the patron saint of skaters.

The sport also became fashionable in France: Marie Antoinette and Napoleon both enjoyed skating.

Skating in Brooklyn

The first properly maintained and organised rink in America was in New York's Central Park. In 1862, two years after its formation, the New York Skating Club organised the first skating carnival on the frozen Union Pond in Brooklyn.

The first mechanically refrigerated rink opened in Britain in 1876. It was private, situated near the King's Road, Chelsea, with an ice surface measuring 40 feet by 24 feet. The National Skating Association of Britain was instituted in 1879 – the first of its kind.

ST. LYDWINA OF HOLLAND *broke a rib – and became the patron saint of skaters in 1396. Skating is thought to have begun in Scandinavia 3,000 years ago.*

RIGHT: THE FIRST RINK *in America was built at Central Park, New York, about 1862.*

In This Field
July 25 1831
Will be tried a new
PATENT GRAIN CUTTER
WORKED BY HORSE POWER
invented by
C. H. McCORMICK

Part 9

AGRICULTURE

The horse has given way to the tractor, the reaper and the thresher to the harvester, the milkmaid to the machine and the sower to the seed-drill. Today's farmhand is a mechanic more than anything else.

Not without protest! Not only the farmers shook their heads when the traditional methods began to disappear: they resisted change because they did not trust new-fangled ways of doing jobs that had been done in the old way for centuries; their labourers resisted because they feared far more for the disappearance of their jobs altogether – they went on strike when the seed-drill first appeared and they set fire to the first reaping machines. And in many cases their fears were well-founded. Most of the agricultural implements in use today are nothing more than mechanical means of doing those traditional tasks, but they are faster, they are more thorough and they certainly are more productive. Though they may have drastically changed the life of the farmer and the labourer, they have made it possible to grow richer crops and make use of land that might otherwise have been wasted or abandoned.

THE HARVESTING MACHINE. *In 1848, the Californian Gold Rush drew away so many men from the farms of America that the harvest could not be brought in. Three years previously, Cyrus McCormack, of Virginia, had perfected the design of a mechanical grain-cutter, invented by his father, Robert. Now, with the farmers crying out for help, Cyrus McCormack's machines sold in huge quantities . . . virtually worth their weight in gold.*

The plough

THE ROTHERHAM PLOUGH has good claim to being the first plough in Britain to go into factory production and become widely established as a uniform design in use throughout the country.

Of course, the plough is one of the earliest agricultural implements and has appeared in various forms since man first began to till the soil for a living. In 2,000 B.C. the early Egyptians were using a form of plough that was something like a pickaxe, gripped in both hands; spade-like ploughs, called 'cashcroms', operated as a lever by the hand and foot, were used by the ancient Britons; in many places the early plough was little more than a digging stick.

Later developments followed. Digging points were fitted with an ox horn. As the use of certain metals was discovered, points were covered with flint or bronze or iron. But these early implements only stirred the surface of the earth and were not able to turn a furrow slice.

Single furrow

With the coming of the Angles and Saxons to Britain, a heavy, single furrow, wooden plough was introduced. Later, the 'Kent' plough, which we can read about in a '*Boke of Husbandrie*' published in 1523, was able to turn the furrow slice either to the left or right, so that successive furrows could all be the same way. A little over a hundred years after that, Walter Blith, an agricultural writer, laid down the essentials of good plough design in another book.

THREE FURROWS AT ONCE. *The Ivel tractor was the first real "mechanical horse". In use from 1902 until the First World War which brought more modern designs.*

It was shortly after this that the Rotherham plough appeared. It was imported originally from Holland, although a man called Joseph Foljambe patented a plough of the same type in 1730. One authority on the history of the plough suggests that Foljambe built the first Rotherham plough under the direction of Walter Blith himself. But this seems unlikely.

Patent challenged

Joseph Foljambe sold his patent to Disney Stayforth, who then manufactured the plough at Rotherham – from where it got its name. Later the patent was challenged and set aside on the grounds that the plough was only an improvement, not an invention.

There are plenty of other stories about the origins of the plough, but they all end up at Rotherham, and whatever the disagreements about the plough's history, there is no doubt that it *was* a great improvement and seems to be the first plough to be manufactured on any great scale to a uniform design.

The mould board (the board that turns the furrow) was curved, not flat, in order to follow the turn of the furrow slice. Also, the design provided strength, at the same time reducing the overall weight of the plough. The Rotherham plough was made mostly of wood, with draught irons, coulter (the vertical blade in front

RESPECTED BY CATTLE. *Barbed wire changed the face of the American plains in the 1880s. Grazing cattle could at last be controlled.*

of the share), share (the main horizontal cutting blade) and sole (the bottom part of the plough) all of iron. The wooden mouldboard was covered with iron plates to reduce the wear on it.

The subsoil plough

IN 1823, JAMES SMITH, an enthusiastic farmer, took over a farm in Dainston. The land was marshy and needed draining. Smith's friends were doubtful whether anything could be done to improve the land and predicted that he would find it too hard a task.

Smith was not deterred. He designed a plough that would stir up and break the soil just beneath the surface without disturbing the surface itself too much. This enabled the water to escape more easily through the soil to the drains. It also let in air which dried out the soil still more.

Farm workers complained of the weight of the plough, which needed six horses to draw it. But Smith had already discovered that any lighter plough jumped out of the soil almost at once. His experiments worked well and neighbouring farmers were impressed, despite their initial pessimism, and adopted the invention on their own lands.

There had been one or two earlier but never very successful experiments in

How Farmer Smith conquered the marsh-land

draining ploughs. During the 1760s Mr. Knowles and Mr. Cuthbert Clarke both received prizes from the Royal Society of Arts for constructing ploughs specifically intended for draining the land. Both were equally heavy and cumbersome – far heavier even than Smith's subsoil plough. Cuthbert's, for instance, excavated a trench 12 inches deep and 18 inches wide and needed 20 horses to draw it along.

The steam plough

STEAM POWER WAS QUICKLY adapted to all kinds of uses: ploughing was one of the first. During the first half of the nineteenth century, a great deal of attention was paid to the use of steam ploughs and various systems were designed.

So many experiments were in fact taking place that it is impossible to pick on any one as the very first. But one of the earliest successes was a single-engined system using a movable anchor and rope porters to draw the plough across the field. This was designed by John Fowler and awarded a prize of £500 by the Royal Agricultural Society of England in 1858.

Later, the movable anchor was replaced by a second engine. The engines moved steadily forward along either side of the field and the plough was dragged to and fro across the field between them. The rope porters that were attached to the plough were winched in and released by each engine in turn.

One of the great advantages of steam ploughing was that up to ten furrows could be ploughed at once. Also, the heavy engine provided a firm anchor for ploughing very heavy or very wet soils which it would have been difficult for horses to plough and which would present problems even for a modern medium powered tractor.

Ten furrows

Some engines developed 100 horsepower and could exert a draught (pull) of four tons at a speed of four miles an hour. With a single working unit of two engines and one plough, 30 to 50 acres could be ploughed efficiently in one day.

The disadvantages of steam ploughing included the time it took to raise steam and the liability of the heavy engines to get bogged down in soft ground. Also, steam ploughing gear was not economical for small fields.

The farm tractor

THE TRACTOR WAS DEVELOPED to replace the horse, which for centuries had done all the work of hauling and carting around the farm.

Massive steam engines at first took away some of the work of ploughing. For some time before the tractor was invented these cumbersome monsters were used to drive winches to which ploughs were attached by cables and dragged across the fields. But the steam engines could not be used very easily for any other farmwork.

The first real tractors, capable of doing

MEDALS FOR ENDEAVOUR. *James Smith's reaping machine was not a success, but it paved the way for more sophisticated machines and earned its inventor several awards.*

a variety of light and heavy work, were small and powered by an internal combustion engine, and the first of these mechanical horses was the Ivel tractor.

One speed forward

The first design appeared in 1902 and sold steadily until an improved type appeared during the First World War. There was only one speed forward and one in reverse. Like the Cugnot steam tractor of 1770 (which had been intended for towing artillery only), the Ivel tractor had only three wheels. As a result, it was very light in front and had a dangerous and nasty tendency to overturn when pulling heavy loads or if it hit an obstacle. However it could plough up to three furrows at once in soil that was not too heavy.

The Ivel was designed by Dan Albone of Biggleswade in Bedfordshire. He started experimenting in 1897 and by 1903 had formed a company with a London office and a factory in his home town.

In 1917, 5,000 tractors were ordered and delivered from the Ford works in America in order to help combat wartime food shortages. It was as a result of their widespread use and success in increasing productivity that the farming community was finally convinced of the value of mechanisation.

The seed drill

JETHRO TULL INTRODUCED THE first successful seed drill in 1701. It was some time before his idea caught on, but it was of enormous importance. Before then, land had never been used to its maximum efficiency. Tull invented the drill in order to increase food production and make the best use of the land available.

He began his experiments at a time when many people were drifting away from the land in order to go and live in the growing industrial towns. There were therefore less people to work the land. Also, there was an increase in the price of seed. As a result there were strong inducements to increase the yield and reduce the labour in growing wheat.

Seed in rows

Previously, seed had been sown largely by hand. Tull's machine made the channels, sowed the seed and covered it all in one operation. The channels were each one foot distant from the next. One bushel of seed was sown per acre at a cost of six pence. By sowing the seed in rows instead of broadcast (scattered freely by hand), the soil around the growing plants could be weeded and aerated more easily.

The idea for the drill came from parts of an organ, a wheelbarrow and a cydermill. Between the two big front wheels was the hopper for the centre coulter (the knife that digs the channel); the two outside coulters were fed from the two hoppers held between the smaller rear wheels. All three coulters had a channel grooved down their backs to guide the seed into the soil. The rate at which the seed left the hoppers was controlled through notched cylinders by a thin metal plate and spring.

Workers fired

The drill met a good deal of opposition from those of Tull's farm labourers who had stayed on the land. They saw it doing them out of a job and in protest they went on strike. Tull dismissed them immediately and quickly perfected his machine in order to replace them.

Although some Scottish farmers were enthusiastic, the drill did not travel far beyond Tull's own farm. It was not until 1800 – almost a hundred years later – that drilling became widely practised in England, when two Suffolk drillmakers, Smyth of Peasenhall and Garrett of Leiston helped to popularise it with their own designs. Smyth sent travellers about the country who offered to drill seed at 2/6 an acre.

Of course, there had been much earlier attempts to produce a drill. As early as 2500 B.C., the Chinese had used a kind of wheelbarrow, and there is evidence that the Babylonians used a form of drill plough a thousand years later.

The 'dibble'

In Britain, interest in drilling did not begin until the seventeenth century. In 1600, Sir Hugh Platt recommended a method of dibbling wheat (a 'dibble' is an instrument for making holes in the ground for seeds), in his book 'The Setting of Corn'.

One other experiment that fell through was designed by John Worlidge, in 1669. He tested the idea of a seed-dropping device and outlined a machine for the purpose. However, when this was constructed, it failed to work.

The reaping machine

BELL'S REAPING MACHINE OF 1826 was not the first to be invented but it was the first to work with any degree of reliability.

The elder Pliny mentioned the use of a mechanised reaper in the Roman Empire, in the First century A.D. This was a two-wheeled cart-like vehicle which was pushed into a standing crop of wheat by a horse or ox. The iron teeth at the front of the machine beheaded the corn and the ears fell into the cart.

Salmon of Woburn was probably the first to apply the cutting action of shears to a reaping machine when he designed a reaper in 1807. It was not very efficient.

Five years later, James Smith invented a reaping machine that he entered for a £600 prize offered by Dalkeith Young Farmers' Club. He failed to win the prize. The following year he tried an improved machine but had trouble with the rough ground. It is reported that the reaper *'fell into a sudden hollow, the cutter stuck fast and part of the mechanism was broken'*. However, Smith was awarded a piece of plate worth 50 guineas for *'meritorious endeavour'* and later received another plate from the Highland Society.

Russian medal

Even the Czar of Russia became interested when he saw a model of the reaper in St. Petersburg. He sent a gold medal to James Smith.

Between 1800 and 1826 there were several other attempts to produce a successful reaper but all failed. Consequently, when it came to testing out his own machine, Patrick Bell was extremely cautious, in case that, too, failed to work. He made his first trial in a barn with a crop that he had planted by hand, stalk by stalk. When this was successful, he made his next trial outdoors – by night.

His reaper was pushed from behind by two horses. Revolving sails held the corn to the clipping knives and a canvas drum delivered it in a swath alongside the reaper to await collection. It is recorded that the reaper cut on average one acre of wheat an hour.

Limited success

Bell refused to patent his machine even though his work on it had left him with very little money. As a result, he could not afford good materials or good workmanship on the reaper and its immediate success was only limited. Furthermore, no big firm was prepared to manufacture it in quantity. This was because the big firms were afraid of the popular opposition to farm machinery of any kind in England at that time.

In America the situation was different. Some of Bell's reapers were taken over there and Bell himself went there in 1832 to give free advice. The huge continent was waiting to be exploited. And it was there, the year before Bell's visit, that the McCormick reaper appeared – the first mechanical reaper to gain popular acceptance.

It was Cyrus McCormick, of Virginia, who got most of the credit for the invention, although his father, Robert, had been working on a design for years. But it was Cyrus's drive and ambition that was responsible for the astounding success of the reaper and particularly for the success of the improved version, which appeared in 1845.

The gold rush

Four years later the Californian Gold Rush syphoned off a great many farm labourers, leaving the land without enough men to bring in the harvest. Sales of the labour-saving reaper shot up.

Twenty years after the McCormick reaper first appeared it was introduced for the first time into England, at the Great Exhibition in London. The interest it aroused, together with public interest in another machine from America, the Hussey reaper, led to the general acceptance of mechanical reaping in England.

One last point is interesting: both McCormick and Hussey later proved that Bell's original reaper was equal in ability to their own when it was properly made, with the right materials and the necessary precision.

The threshing machine

ANDREW MEIKLE LEARNT FROM HIS FATHER, James, how to make a rotary beating machine that copied the action of a flail – until then the traditional method of threshing, or 'thrashing', corn. Andrew's improved machine appeared about 1786–8.

A stout wooden frame turned on a roller. Strips of strong cloth or leather were tied to the frame. As the roller turned, and the strips beat the corn, an artificial wind blew the ears and chaff away from the grain. At first, horses were used to drive the machine; later, the horses were replaced by steam power.

The earliest method of threshing corn that is recorded was to drive animals – oxen, mules or horses – over the sheaves of corn that were laid out in a circle on the threshing floor. The Egyptians used this method. An alternative was to beat handfuls of corn against a stone.

Iron teeth

The Chinese used rollers driven round over the corn. The Romans used a heavy wooden sleigh with flints or iron teeth fixed to the underside. But for centuries in most areas of the world the flail was the most popular instrument and was used well into the last century.

Threshing with the flail meant constant winter work for farm labourers when there was little else to do. For one thing, it saved them from having to fall back on parish relief for support when they were out of a job. Parish relief was undignified and unpleasant.

Security threatened

Andrew Meikle's thresher clearly threatened the security of this work. Labourers everywhere protested strongly and riots broke out, aimed against the machine, in Kent, in August, 1830. These became known as the Captain Swing riots. Rather than risking their barns being burnt down by angry labourers, many farmers moved their threshers out into their yards for the men to destroy more easily.

The riots were suppressed ruthlessly by the landowners but nonetheless the spread of the machine was slowed down a great deal. In Cambridgeshire, for instance, the flail was used for a long time as the 'poverty stick' – farmers and parish officers together provided the unemployed with work, flailing in the barn.

The firm of Richard Garrett of Leiston, in Suffolk, was one of the first to popularise the thresher in England, during the middle of the last century.

The combine harvester

DURING THE LAST CENTURY there were a great many attempts to combine the functions of the reaper and the thresher into one machine. Two harvesters were patented in America in 1836 and seven years later a more successful harvester was put together by John Ridley, an Australian farmer.

Harvesting giants

However, most of the harvesting giants that appeared in the American mid-west and California during the middle of the century required teams of between 30 and 40 mules to pull them – not very efficient and certainly not very easy to handle, as can be imagined.

The forerunner of the modern combine harvester was made by another Australian, Hugh McKay, in 1884. The Massey-Harris Company adopted the machine, developed and improved it, and in 1910 introduced the first really practical mechanical Reaper-Thresher.

The first combine harvester to appear in England was assembled in London in in 1928.

Breeding

A GREAT MANY PEOPLE THINK that a particular breed of animal has always existed just as we know it today. This is not true. Animals are adapted through breeding to meet the needs of people at a certain time. For instance, one and a half centuries ago there was very little fat available in England, so breeds of cattle were improved specifically to increase their fat yield.

The first person to investigate seriously the possibility of improving breeds of cattle was Robert Bakewell. And the first

breed to be improved by him, in 1750, was the Longhorn. The breed is now almost extinct.

Years of study

In Bakewell's day, Britain was about the only country in which there were enough wealthy people who wanted to pay for high-quality meat, the sort of people who were no longer prepared to eat the product from an old cow, an ox that had spent years in the plough team or a sheep that had borne three or four fleeces already. So Bakewell tried to meet the new demand and after years of study and experimentation succeeded in creating an established breed that could yield at an early age carcasses of meat as fat as the public might want.

Bakewell was most meticulous as to the qualities he looked for in improved breeds. One of the first improved rams that he produced was the Leicester ram Twopounder.

'The collar of a ram,' wrote Bakewell, *should be thick and bowed like a swan so that the drops from his nose may fall on his breast; that he should have an eye like a hawk and a heel like a lark. The head long and thin between the eyes . . . Nothing but first rate loins, thighs and scrags can support in and in breeding*'.

'Strong loin, strong constitution,' said one old farmer from whom Bakewell took advice.

One effect of Bakewell's work on breeding is that over a large part of the world, especially in the new countries, beef, mutton and lamb have come mostly from breeds originally created in Britain. So, New Zealand lamb originated from the Romney Ewe and beef in North America is partly derived from breeds created in Hereford, North East Scotland and Devon.

The cream separator

THE FIRST SUCCESSFUL MACHINE for separating cream from milk was invented in 1878 by a Swedish chemist, Dr Gustave de Laval. In the next year, de Laval's cream separator was shown in England at the Kilburn Royal Show. It was awarded a silver medal.

It had never been very difficult to separate cream from milk but it used to take a long time. Until the end of the last century the most common method used was to leave the milk to stand for 24 hours during which time the cream

formed a layer on the surface. It was then skimmed off. This was known as the gravitational method.

Centrifugal force

De Laval's separator used centrifugal force instead—a simple but effective breakthrough. There are still a few machines of his original design. The milk is fed into the bowl and the cream is separated from the relatively heavier milk as the drum spins round.

Thirty gallons of milk an hour could be passed through the separator, which was powered either by horse or steam. However, there was altogether a great deal too much equipment to fit into the average-size dairy. Most of Laval's separators were used in factories.

The milking machine

THIS EARLY MILKING MACHINE was the first practical one to be produced. It was invented by a sanitary engineer called William Murchland, in 1889. There had been several experimental machines before but most were either unhygienic or tended to injure the cow.

The *Illustrated London News* had some sad words to say in 1862 about one still earlier machine:

'*We seem quickly to be losing the poetry of rustic life. The mower is no longer required at the scythe, nor the reaper at the sickle; that bent figure at the barn-*

LANCASTER LONGHORN. *Cows haven't always been the same. Robert Bakewell's breeding experiments completely changed the longhorn in 1750. But even this variety is now virtually extinct.*

door swinging to the music of the flail is gone; the whistle of the ploughboy is gradually dying in the distance; and now we are called upon to dismiss the ruddy milkmaid.'

The article ended on a more optimistic note:

'*While the operation* (of the machine) *is distressingly practical, it is cleanly, and, we believe, agreeable to the cow. The milk is withdrawn at the rate of a gallon a minute.*'

Hand pump

Murchland's milking machine was not quite so fast, but it was probably a great deal more efficient. One cow could be milked in about eight to ten minutes. The machine worked on the principle of continuous suction. Air was drawn in by means of a hand pump from an iron pipe fitted round the milking shed. This formed a vacuum. The teats were attached to the cow's udder and the milk was sucked down through the teatcups.

There was, however, one major drawback in Murchland's machine, as with all the early experiments. Calves normally suck with a pulsating motion, not continuously. Continuous suction eventually

makes the udder extremely sore. It was not until 1895 that a machine with a pulsating action was produced.

The lawnmower

GRASS HAD ALWAYS BEEN CUT by scythe until Edwin Budding invented his mechanical lawnmower in 1830. Normally, for scything, grass had to be a little damp. Budding's lawnmower would cut it dry, though he did admit that a second run might be necessary if the grass was tall.

In his advertisement of the lawnmower's many advantages, he said proudly that;

'*Grass too weak to stand against a scythe . . . may be cut by machine as required, and the eye will never be offended by those circular scars, inequalities and bare places so commonly made by the best mowers with the scythe . . . Country gentlemen may find in my machine themselves an amusing, useful and healthy exercise.*'

Budding was an engineer in a textile factory. The principle on which his mower worked – a rotating cutter operating against a fixed knife – was a clever adaptation from the method used in the factory of shearing the nap, or loose ends, from cloth.

If the gentlemen who bought the lawnmower wanted a great deal of exercise, they could push it from behind. But in case this proved too hard another handle was provided in front so that a second person could lend a hand by pulling.

The first mechanised lawnmower was built in 1893. It was a huge Leyland steam mower specially built for large estates.

The first mechanical lawnmower that could be used in an average-sized garden was invented by Ransomes of Ipswich, in 1905. It was driven by an internal combustion engine similar to that used by early cars.

Barbed wire

IT IS HARD TO TELL when exactly barbed wire was invented. Obviously it would not have been very difficult for almost anyone to have the idea of twisting short strands at intervals into a length of wire, though several people – one of them was called Lucien Smith – claimed that they had patented the idea.

There was certainly a great need for barbed wire on the wide plains of North America where as much money had to be spent on fencing as on cattle. It was impossible to grow hedges over those great distances; stone walls were unthinkable; wooden fences rotted and were easily broken. But barbed wire could be transported without much trouble, put up quickly, and was respected by the cattle. What was needed was a reliable machine to manufacture it by the mile.

New type

This machine was first produced in 1873, after a 60-year-old rancher, Joseph Glidden, together with his friend, Jacob Haish, paid a visit to the county fair of De Kalb, Illinois. There they saw a new type of barbed wire which was exhibited by Henry Rose.

Both men were greatly impressed. Without saying anything to each other about their interest, on returning to their separate homes each worked out a machine for threading in the spurs automatically to a length of wire.

Glidden got his patent out first. Haish challenged him, saying that the idea was not original, but he lost his case and although he started producing his own wire a few years later it was Joseph Glidden who reaped both the reputation and the wealth from the manufacture of 'his' barbed wire.

Mass-produced barbed wire made life a good deal easier for the settlers and

gave them a greater sense of security, but there were those who were not so pleased with the invention. These were the cattle barons who had been used to driving their enormous herds over any land that lay in their path. Not only did they try to tear down the wire but often they hired men to shoot anyone they saw attempting to put the wire up.

ABOVE. HEALTHY EXERCISE. *With Edwin Budding's lawnmower, country gentlemen could both exercise and keep their lawns neatly cropped.*

BELOW. JOHN WORLIDGE'S SEED-DROPPER *preceded Jethro Tull's successful seed drill of 1701 by two years.*

Part 10

INDUSTRIAL

*We associate a wide variety of materials
and machines with the industrial
world and we tend to think of 'industry'
as something new. But by now you will
probably have seen that some of the
'newest' things are sometimes the
oldest and some that appear old are
really quite new.*

*In industry it is just the same. For
instance the Egyptians used glass for
ornaments 4,500 years ago; a wooden
locking device for a door almost as old
has recently been discovered; paper,
on the other hand, did not appear
until 100 years after the birth of Christ,
though we might think it had been
around for long before that.*

*The calculating machine is as old as the
abacus and the Aztecs of South America
were making rubber balls for games at
least 800 years ago. If none of those
facts surprise you, one way or the
other, perhaps these will: the first
automatic assembly line was used in a
Royal Navy biscuit factory and
electricity was first called by that name
600 years before Christ.*

1913. THE FIRST ASSEMBLY LINE *in the motor industry.
This was Henry Ford's idea. He split up the assembly of his
famous cars into 84 stages. By doing so, he reduced the number
of workers needed by one third.*

Automatic control

HUMPHREY POTTER WAS EMPLOYED at a salary of one shilling a week in a coal mine in Warwickshire, about 250 years ago. His job was to open two valves, one after the other, which let steam into and out of a steam engine used for pumping water out of the mine.

But Humphrey Potter was also trying to study mathematics and the constant bother of laying down his pen and paper in order to open and close the valves disturbed his concentration on what he thought was a far more important subject.

So he attached some cords from the handles on the valves to the ends of the engine's rocking beam, which rocked up and down like a see-saw across the top of the engine. The beam pulled the cords, first one, then the other, as it rocked; the valves were opened, first one, then the other; and Humphrey Potter was free to concentrate on his maths.

From such a simple invention developed automatic control in industry, now used in most fundamental mechanical processes of production.

The assembly line

THE ORIGINS OF THE ASSEMBLY LINE are very curious. The first assembly line was almost certainly that used in a Royal Naval biscuit-making factory in 1833!

A separate and specialised machine was used for kneading the dough, but the work of shaping and cutting the biscuits was then passed from man to man on trays carried round automatically on power-driven rollers.

An equally unusual task was carried out by another early version of the assembly line in a pig factory in Cincinatti, in the United States, thirty years later. Carcasses of the dead pigs travelled through the factory on an overhead conveyor belt. Each worker had to make one cut from the carcass as it passed him.

Commercial success

Strange beginnings for an industrial system that has come to be used with tremendous commercial success in important industries such as motor manufacturing and engineering!

The first assembly line to be used in the motor industry was installed by Henry Ford in 1913. He split up the assembly of his famous cars into 84 stages and thereby

reduced the number of workers he needed by one third.

The chassis of the vehicle to be worked on was placed on rails and towed past a series of work stations at each of which further stages of the assembly were completed.

The miner's safety lamp

THE LIFE OF COAL MINERS had always been dirty and exhausting and dangerous. It became increasingly dangerous when deeper pits were sunk in Britain in the late eighteenth and early nineteenth centuries. Sudden escapes of gas from the coal seams were ignited by the miners' naked candles – their only light – causing terrible explosions. Hundreds of miners lost their lives.

The need for better, safer lights was obvious but before Sir Humphry Davy produced his 'Davy' lamp the only real attempt to solve the problem was in about 1811 when a Dr. Clanny suggested a device that insulated the flame of the candle inside a glass cylinder.

This was not very satisfactory nor very popular. It was left to Davy to convince both the miners and the owners of the coal mines that safety lamps should be adopted in both their interests. 'Davy' lamps were based on the principle that if a flame was surrounded by a wire gauze of fine mesh it would be sufficiently insulated to avoid any disastrous contact with underground gases.

Prototypes of these lamps were successfully demonstrated at Hebburn Colliery,

ABOVE. ASSEMBLY LINE. *In 1913 Henry Ford brought mechanisation to the motor industry and revolutionised the process of manufacture.*

RIGHT. FIRST WORKSHOP. *By contrast, Ford had begun producing cars by hand in his barn.*

a notoriously dangerous and 'fiery' mine in Durham, on January 9th, 1816.

Even so, it took more than 30 years before 'Davy' lamps were used widely. Miners continually protested that the light given out by the new lamps was much dimmer than their old naked candles. Some Midlands miners claimed extra money because they were unable to work so quickly or efficiently with the new lamps. And they were paid, too.

Electricity

IT WOULD BE ABSURD TO TALK about 'the first electricity'. There is electricity all around us, even though we cannot see it or hear it. But we have learnt to recognise it and harness it to work for us.

The first man to notice the phenomenon of electricity was a Greek philosopher called Thales of Miletos, in 600 B.C. When he held a piece of amber and rubbed it, Thales noticed that it became capable of attracting small bits of straw. The Greek word for amber is *elektron*, so Thales called the invisible force 'electricity'.

The first machine to demonstrate electricity clearly to the naked eye was invented a long time later, in the seven-

teenth century, by the Burgomaster of Magdeburg, in Germany. This burgomaster, Otto van Guericke, caused sparks to leap across the gap between two metal balls.

Blacksmith's son

However, electricity was not developed as a source of power and energy directly to help man until two centuries later, when Michael Faraday, one of Britain's most eminent scientists and inventors of the nineteenth century, carried out a series of major experiments. Faraday was the son of a blacksmith, but he rose to become assistant to Sir Humphry Davy at the Royal Institution off Piccadilly, which is where he did most of his work.

Faraday's greatest discovery was how to produce an electric current at will. He realised that a stationary magnet does not produce electricity just by itself. But when he moved the magnet through a coil of wire a current of electricity began

to flow inside the coil.

This is called the technique of electromagnetic induction – inducing an electric current – and this knowledge formed the basis of electrical engineering on which our present use of electricity as an important source of power is based.

The cash register

A SHIP'S LOG LED TO the invention of the world's first cash register.

James Ritty owned a drapery store in Dayton, Ohio. For some time he had suspected some illicit 'pocketing' of cash by his shop assistants. So Ritty looked around for some way to put a stop to this.

It was on the return voyage from a business trip to Europe, in 1879, that he took notice of the ship's log that recorded the distance the ship travelled each day. Perhaps, thought Ritty, he could make use of the idea.

Together with his brother, James, also a keen amateur mechanic, Ritty adapted the principle of the ship's log to a machine that would register the amount of money taken each time by his assistants and also the total amount of money taken each day.

Clock face

The first cash register was patented by Ritty on November 4th, 1879. It had two rows of numbered keys and a clock face, or dial, with concentric rings marked in dollars and cents. When the keys were pressed, the amount of money taken in any particular sale was indicated on the dial and recorded by a printing device that was attached to the machine.

In later models the dial was replaced by a glass case on top of the machine, in which flags sprung up to show the amount of the sale, which could then be immediately seen by the customer or by a nearby manager.

Nylon

NYLON WAS THE FIRST truly synthetic fibre. It was discovered only by a concentrated, expensive research and development programme launched in the United States in 1927. The first nylon stockings for women appeared in April, ten years later.

The American company, E.I. du Pont de Nemours & Co., agreed to give financial backing to the research programme headed by a young Harvard chemist called Wallace H. Carothers.

In 1930, three years after the beginning of the research programme, Dr. Julian Hill, a member of Carothers' team, while trying to extract a sample of heated polymer (a compound of substances) from a vessel, noticed that the molten material could be stretched to form a long fibre. Dr. Hill also noticed that this particular material could be drawn out even longer when it cooled, and that the process of stretching the material greatly increased its strength and elasticity.

Slow progress

Despite this discovery, neither Carothers nor the du Pont company could see how they could obtain any great commercial value out of this. They very nearly abandoned their work on the development of synthetic fibre completely. Reluctantly, Carothers was persuaded to pursue the research. Progress continued slowly during the 1930s and success was achieved finally in 1937.

The programme had taken over ten years and cost well over 20 million dollars but the eventual commercial success of nylon and other synthetic fibres was enormous.

The electric light bulb

OF ALL THE INVENTIONS of the nineteenth century, one of the most simple and yet most useful is now taken for granted – the electric light bulb, invented by Thomas Edison in 1879.

Other men had experimented with electricity long before Edison made his revolutionary discovery. More than a century earlier, Benjamin Franklin had tried to store the electricity caused when his kite was struck by lightning. Sir Humphry Davy was another pioneer, so was the Italian scientist, Galvani.

Edison knew of the work of these men, and was determined, if possible, to use electricity to provide some form of long-lasting lighting. The arc light had been invented earlier – though it was not exactly long-lasting. Edison conducted numerous experiments into glass bulbs including, on one memorable occasion, a red hair, which he tore on impulse from the beard of a visiting friend!

Cellophane

THE INVENTION OF CELLOPHANE provides a classic example of the maxim: '*If at first you don't succeed, try, try, try again*'.

Jacques Brandenburger, a Swiss-born French chemist, did not succeed at first but is generally credited with the pioneer work that made cellophane possible. Brandenburger was interested in the packaging industry, which was developing very fast, and he was anxious to produce a thin, transparent covering that could be used for packaging.

He made various experiments between 1900 and 1912 with a variety of materials and had several failures. Most of his results were too thick and not transparent enough. But in 1912 he succeeded, by adding glycol (originally the word was meant as a name for a substance somewhere between glycerine and alcohol) to a thin sheet of film cellulose. Brandenberger patented the film and the machinery with which he had made it and began to manufacture the material in the United States and in Europe.

A year later, a French company, Comtoir de Textiles Artificiales, agreed to finance Brandenberger's project and fourteen years after he had first begun work on the idea cellophane was being manufactured commercially for the first time.

Atomic energy

ATOMIC ENERGY IS THE greatest potential source of energy available to the world. It will become more and more important as natural fuels such as coal, oil and natural gas exhaust themselves in the centuries to come.

But the cost and complexity of harnessing atomic energy to these peaceful purposes is enormous and in many people's minds atomic energy is linked not with use in peacetime but with terrible weapons of war.

It was in war that atomic energy was first used, in the form of the atom bomb, developed through a £500 million research and development programme and dropped on the Japanese cities of Hiroshima and Nagasaki in August, 1945.

The discovery of radio-activity, which began the process of inquiry into atomic energy, was made in 1896, but the 'nuclear age' did not really begin until 1939, when Otto Frisch, an Austrian-born English physicist, and Lise Meitner, a German scientist, discovered what we call 'nuclear fission'.

Nucleus splits

What they discovered sounds complicated but the result was alarmingly straightforward. They found that when a particular kind of sub-atomic particle – a neutron – entered the nucleus of a uranium atom then the nucleus splits into two roughly equal parts. In splitting like this it releases an enormous amount of energy and at the same time produces more neutrons that can be used to split more uranium nuclei.

This is a chain reaction and it builds up very quickly to a gigantic release of energy. As the scale of the Second World War increased, scientists in Britain and America were soon able to convince their leaders that with the help of nuclear fission they could create bombs many thousands of times more powerful than anything that had so far been developed.

The result was the atom bomb.

Back from America

At the end of the war, British scientists returning from America began to develop nuclear energy independently. By 1950 it had become clear to them that it could be harnessed to produce electricity on a large, commercial scale. A power programme aimed at achieving this object was announced by the British government five years later.

The cotton gin

COTTON BECAME A BIG INDUSTRY when Eli Whitney invented his famous 'gin'.

Although cotton crops had been harvested in the 'Deep South' of America since the middle of the eighteenth century, by the end of the century cotton was still being picked and cleaned with laborious slowness by negro slaves. One slave working all day could not clean much more than one pound of cotton.

By 1792, when Eli Whitney, a dedicated

and inventive engineer, visited a cotton planter friend in Virginia, only about 140,000 pounds of cotton were being produced in America each year.

Whitney was amazed at this inefficiency and brooded on the problem of how to separate the soft, white fibres from the hard, black seeds – the job the negroes had to do by hand – with greater ease.

Walking in the country one day, he found the answer. He saw a cat creeping toward a fence set up round a chicken run. There was no gap wide enough for the cat. All it could do was to push a paw through the fence and wait for a chicken to come close. When one did, the cat grabbed at its leg and tried to draw it near enough to kill.

Whitney shooed away the cat, so the story goes, but went away with the idea he needed. He withdrew to a makeshift workshop and produced his 'cotton gin' within three days.

The 'gin' consisted of a revolving cylinder with hundreds of sharp wire hooks mounted on it and arranged rather like the teeth of several circular saws (at least as sharp as the claws of the cat). The cylinder was set behind a breastwork which had grooves in it lined up with the cylinder's hooks – like the claws of the cat stretching through the slit in the fence.

Once a cotton 'boll' was swept up by the hooks, the hard seeds were caught in the breastwork and the cotton fibres were dragged through the grooves. The seeds then fell into a hopper while rotating brushes brushed off the fibres from the hooks. Power for turning the cylinder was provided either by slaves turning it by hand, or by water power or horses.

Whitney patented the machine in 1794 but too late: it was immediately copied by hundreds of amateur engineers. By the early years of the next century, cotton production – in answer to the great demand in Europe and Britain – had gone up to 85 million pounds a year by 1810. Quite a step forward!

The addressograph

THE FIRST MANUFACTURED ADDRESSO-GRAPH was called the 'Baby O' – a name that is not easy to forget. It was invented by an American, Joseph Duncan, of Sioux City, Iowa, and was first sold to firms in Chicago in July, 1893. Three years later, Duncan took out a patent

SELLING SYNTHETIC FIBRES. *A giant leg, 35 feet high and weighing two tons, was used to launch nylon stockings on the American public.*

on it.

The addressograph was only one of a series of printing developments during the nineteenth century. Its aim was to save time printing names and addresses where large numbers of letters had to be sent regularly to the same people.

Several unsuccessful attempts had been made earlier to produce a machine, and Duncan himself had created an experimental machine the year before the 'Baby O'. That first machine was never sold

commercially but the 'Baby O' was no more than an improved version of the original.

The original consisted of a hexagonal wooden block on which were glued names and addresses formed out of type torn off rubber stamps. The block was revolved by hand. As it turned a new name and address was inked and printed on to an envelope.

Later addressographs worked mechanically.

Glass

GLASS IS ONE OF THE OLDEST forms of decoration. Even today both primitive tribes and the people of civilised countries take enormous pleasure in the glitter of coloured pieces of glass. The ancient Egyptians used it to imitate precious stones as long ago as 4,000 B.C.

Their methods of making glass were crude and what they produced was not much like the clear glass used for drinking glasses or windows today. By heating sand over a fire they created a fragile, smooth, glistening surface that they could use instead of diamonds or other stones in, for instance, the eyes of statues.

Around the first century B.C., first Syrian and then Roman craftsmen discovered the art of 'blowing' glass. This was done by mouth through narrow metal tubes before the glass had cooled. And in this way the glass could be distorted into shapes such as drinking vessels, vases and more elaborate forms of ornaments.

Isolated examples

The craft of 'blowing' glass continued for more than 3,000 years. It was still used widely in Britain in the last century, though now there are only isolated examples of the art in existence.

Luxury glass-making was brought to England about 400 years ago by Huguenot craftsmen from Lorraine, in France, who first settled in the Weald of Kent and took their skills around the country to places such as Stourbridge, in Worcestershire, still a major glass-making centre.

Glass was used for windows in the earliest cathedrals. Separate pieces were stained into different colours and assembled painstakingly, piece by piece. But the development of modern plate glass began in Britain only about 150 years ago.

At first the glass was cast on to a flat table. When it had gone rigid through cooling it was polished up.

Paper

ANCIENT CIVILISATIONS, including the Egyptians and the Babylonians, used leather and parchment made out of dried animal skins to write on. Even earlier, the Egyptians had used papyrus. Paper, as we know it, did not appear until almost 100 years after the birth of Christ.

A Chinese manufacturer named Tsai-Lun produced the first sheets of paper. He beat his material to a pulp, diluted the pulp in water and then drained off the water through a square mould that he placed in the liquid. The fibres matted together and formed the paper.

Basic raw materials for the earliest paper included wood and waste linen, which was shredded on crude machines with sharp blades mounted on revolving cylinders. Once the sheets of paper were formed they were hung up to dry. Then they were rubbed gently with smooth stones to produce a good finish.

Stevenage mill

Paper-making spread from China, through Samarkand and Morocco, to Italy and eventually, by the end of the fourteenth century, to Germany and Northern Europe.

In Britain the earliest mention of the art of paper-making refers to a paper mill at Stevenage in Hertfordshire, in about 1495. Nearly a century later Queen Elizabeth I granted a licence to a manufacturer named John Spielman to establish a mill at Dartford, in Kent – still an important paper-producing centre.

Mechanical processes that improved the speed and the quality of paper-making were first developed in the late eighteenth century, principally in France.

Locks

THE USE OF LOCKS TO FASTEN doors and secure houses and possessions can be traced back 4,000 years. A wooden locking device was discovered attached to a door from the ancient ruined palace of Khorsabad, near Nineveh, dating from about 2,000 B.C.

We know that wooden locks and keys were also used by the ancient Egyptians. These consisted of a staple, a vertical piece of wood fixed to a doorpost and containing movable pins, plus a cross-piece forming the bolt. The pins in the staple were arranged so that they would fall into corresponding holes in the bolt. The bolt could not be moved then until the pins were raised by the key.

Metal locks were first used in Roman times. The Romans used iron to make the locks and bronze for the keys. The patterns of the pins on surviving Roman keys show that the Romans were the first craftsmen to use obstacles inside the key-hole to prevent any but the correctly

shaped key from opening a particular lock.

Portable lock

Early Chinese and Russian craftsmen produced the portable lock, or padlock. Medieval and Renaissance locks had a few dangerous extras – machinery that shot bullets at intruders who tampered with the lock, or sharp blades that sliced off the ends of their fingers.

The history of modern locks does not begin until the eighteenth century, when French locksmiths first used tumblers as a secondary safety device. The Yale lock did not appear until 1861, when Linus Yale Jr., of the United States, revolutionised the lock industry with his invention of the pin tumbler lock, which was lighter, smaller and cheaper than previous locks and could be mass produced by mechanical means.

The Windmill

THE DATE OF THE EARLIEST windmill is almost certainly AD 650, when they were built in China and Persia. Windmills were in use in Northern Europe at least 1250 years ago. They are one of the first examples of industrial power and were used mainly for grinding corn and pumping water.

The early Oriental mills had a vertical axle and were known as postmills. Their whole structure was carried on a box-like wooden house mounted on a central vertical post. The structure – house, sails and all – had to be turned by hand so that the sails could face into the wind.

Millers had unusual names for the various parts of the windmill. The brake wheel, next to the sails on the main post, engaged with another wheel called a 'wallower'. This connected with other cogs and gears to operate the millstones themselves that ground the corn.

Water-pumping

There used to be a great many windmills in use, usually on higher ground, where they would catch the wind, in many countries including Britain, where they were used, apart from corn-grinding, to pump water from marshy fenland in Norfolk.

Although most fell out of use or disappeared completely during the industrial revolution of the eighteenth and nineteenth centuries, a few still remained in use even in this century.